*Your attention is called to the new and revised edition of*

ELDERHORST'S MANUAL

OF

# BLOW-PIPE ANALYSIS

AND

# DETERMINATIVE MINERALOGY,

JUST ISSUED.

PHILADELPHIA, APRIL 4, 1866.

The present edition will be found to contain a great deal of additional valuable matter, the result of several years' experience, (since the first edition,) both in the Laboratory and the Lecture Room, which enables us to give a useful Manual for both Teachers and Students.

We will briefly recite some of the additions of special interest to those who may use the blow-pipe as an aid to the study of Mineralogy.

In the Fourth Chapter the Author has given the characteristic physical properties, blow-pipe reactions, and behavior to solvents of all the important Ores of the useful metals, so that a reference to this chapter will at once enable the student or the practical mineralogist and geologist to determine the nature and mineral species of the Ore under examination. But not only the mineral species can thus be easily ascertained; by referring to the methods laid down in the Third Chapter for the detection of metallic oxides, &c., in presence of other compounds, incidental constituents, such as silver in galena, nickel in cobaltine, can be discovered without difficulty. The Sixth Chapter contains Prof. von Kobell's method for the discrimination of minerals by means of the blow-pipe, aided by humid analysis. It is unnecessary to dwell on its merits, as it is almost universally acknowledged to be the best guide for the discrimination of minerals that has ever been published.

The appended tables, containing the behavior of the alkaline earths, the earths proper, and the oxides of the heavy metals before the blow-pipe and to the most important reagents, have been taken from Plattner's work on the blow-pipe, the most thorough treatise on the subject.

We are happy to be able to give below, the opinions of some of our scientific men:

From GEORGE J. BRUSH, Professor of Mineralogy and Metallurgy in Yale College. (*Scientific Department.*)

"The second edition of *Prof. Elderhorst's Manual of Blow-pipe Analysis* will be warmly welcomed by all students of Chemistry and Mineralogy. It fills a want very much felt by both teachers and students, and as the present edition contains not only a thorough treatise on *Quatitative Blow-pipe Analysis*, but also an extended chapter on *Determinative Mineralogy*, this last extracted from Von Kobell's excellent *Tafeln zur Bestimmung der Mineralien.* it forms a most invaluable manual for the Laboratory and Cabinet. I have for several years used the first edition as a text-book for my students, and have found it a most useful and accurate work. I shall most willingly adopt this new and enlarged edition, and I take pleasure in cordially recommending it to all teachers and students of Chemistry and Mineralogy."

From CHARLES A. JOY, PH. D., Professor of Chemistry in Columbia College, New York.

"I am greatly pleased with the improved appearance of this edition. It is now the most convenient work on the blow-pipe we have, and I shall recommend it to all of my friends. All of my pupils who have used it, have made remarkable progress in the detection of metals, and in the determination of minerals."

From Professor JAMES C. BOOTH, (Chemist,) Melter and Refiner, United States Mint.

"After a careful examination of *Elderhorst's Blow-pipe Analysis*, which you have recently published, I am of opinion that it is the best treatise on the subject for the use of beginners and students, and that the experienced chemists will find in it an invaluable vademecum in testing minerals and inorganic bodies generally. I shall therefore recommend it to others and use it myself."

From Dr. WOLCOTT GIBBS, Professor of Chemistry in the Free Academy, New York.

"I most cordially approve of the plan and execution of the treatise, and sincerely hope that its sale may be commensurate with its merits. It is a most welcome and useful addition to our scientific literature."

From Prof. B. HOWARD RAND, M. D., Professor of Chemistry in the Jefferson Medical College, Philadelphia.
Late Professor of Chemistry in the Medical Department, Pennsylvania College;
" " " Franklin Institute;
" " " and Physics in the Philadelphia Central High School.

"I can most cordially commend *Dr. Elderhorst's Manual of Blow-pipe Analysis*. It is clear, concise and accurate, and I shall take pleasure in recommending it to my classes as the best elementary work on the subject."

From Dr. F. A. GENTH.

"CHEMICAL LABORATORY, PHILADELPHIA.

"*Dr. Elderhorst's Manual of Blow-pipe Analysis and Determinative Mineralogy*, just issued by you, is the most convenient work of this kind in the English language, and I shall use it as a text-book for my students, and with great pleasure recommend it most cordially to all my friends who are engaged in chemical and mineralogical pursuits."

From CHARLES P. WILLIAMS, Analytical and Consulting Chemist.

PHILADELPHIA, March 24, 1866.

"I regard *Elderhorst's Manual of Blow-pipe Analysis* as the most convenient elementary text-book in the English language for the student of determinative mineralogy, and as such, for several years have recommended it to students in my laboratory. Though, concise and condensed, no necessary elaboration of detail has been omitted, and it is as useful to the preceptor as to the student. I recommend it with great pleasure."

SCHOOL OF MINES, COLUMBIA COLLEGE, N. Y., March 14, 1866.

THOMAS EGLESTON, Jr., A.M., E.M., Professor of Mineralogy and Metallurgy, says:—"We have always used *Elderhorst's Manual for the Blow-pipe* in preference to any other."

From CHARLES F. CHANDLER, PH. D., Professor of Analytical and Applied Chemistry, and Dean of Faculty, Columbia College, N. Y.

"NEW YORK, March 24, 1866.

"I am very glad to learn that you intend publishing a new edition of *Elderhorst's Blow-pipe Analysis*. As a text-book for students in blow-pipe analysis, and the discrimination of minerals by their chemical reactions, it is far superior to any other book in our language."

From SILAS H. DOUGLASS, Professor of Chemistry and Mineralogy.

"UNIVERSITY OF MICHIGAN, DEPARTMENT OF CHEMISTRY, March 28, 1866.

"I am glad to hear that you are about to issue a new edition of Elderhorst on the Blow-pipe.

"Having used it for several years as a text-book with my classes, I do not hesitate to say that I know of no work capable of taking its place in the Laboratory of Determinative Mineralogy."

From A. B. PRESCOTT, M.D., Asst. Prof. of Chemistry, and Lecturer on Metallurgy.

"CHEMICAL LABORATORY of the UNIVERSITY OF MICHIGAN, March 26, 1866.

"I esteem *Elderhorst's Blowpipe Analysis* as the best Manual extant for the student in Determinative Mineralogy. The systematic tabular methods of Analysis, in the fifth and sixth chapters of the 2d edition, are efficient aids in complete chemical examination. Having been at some inconvenience from its having been recently out of print, I shall gladly welcome a new edition."

From Professor WILLIAM B. RISING, Michigan University.

"FEBRUARY 22, 1866.

"Our school of mines is just starting. We shall use *Elderhorst's Manual*. I have yet to find the man who will say that it is not the standard work for a beginner in blowpiping."

---

*Price of the Manual, sent per mail, prepaid,* $2.50.

Respectfully,

T. ELLWOOD ZELL, Publisher,

17 & 19 South Sixth Street, Philadelphia.

IN PREPARATION.

NEW EDITION

OF

# ELDERHORST'S

# BLOW-PIPE ANALYSIS,

AND

# DETERMINATIVE MINERALOGY.

---

Fourth Edition.

---

REVISED

BY

CHARLES F. CHANDLER, Ph. D.,

PROFESSOR OF ANALYTICAL AND APPLIED CHEMISTRY, SCHOOL OF MINES,

COLUMBIA COLLEGE, NEW YORK.

---

*Ready September, 1867.*

# A MANUAL

OF

# BLOW-PIPE ANALYSIS.

AND

# DETERMINATIVE MINERALOGY.

BY

WILLIAM ELDERHORST, M.D.,
PROFESSOR OF CHEMISTRY IN THE RENSSELAER POLYTECHNIC INSTITUTE.

Third Edition,

---

PHILADELPHIA:
T. ELLWOOD ZELL,
Nos. 17 and 19 SOUTH SIXTH STREET.
NEW YORK:
D. VAN NOSTRAND.
1867.

# PREFACE.

THE present edition of this "Manual" is, like the preceding, designed to serve as a text-book in the instruction in Blowpipe-Analysis and Determinative Mineralogy, in the Rensselaer Polytechnic Institute.

In the first three chapters, but few alterations and additions have been made, fearful of injuring the practical usefulness of the book by an accumulation of too much material. The fourth chapter, containing the characteristics of the most important ores, has been considerably enlarged by increasing the number of species, and by adding an appendix containing the description and blowpipe-reactions of the various kinds of fossil fuel; additions which, I trust, will be especially acceptable to the Mining-Engineer and Geologist. In the selection of the newly added species I have paid particular regard to those occurring in the American Continent; for this reason, many less important ores have found a place in the list to the exclusion of others, which, though more valuable, have not hitherto been found in America.

The fifth chapter, containing a systematic method for the discrimination of inorganic compounds, is a translation, but slightly altered, of the "*Division dichotomique pour reconnaître les minéraux*," as given in Laurent's "*Analyze au Chalumeau.*" It is of no great value to the experienced analyst, but very useful for beginners, and it is on their account that I have given it a place in the Manual.

The sixth chapter is not contained in the first edition. It is hardly necessary to allege any reason for its introduction into this edition. The admirable method of Professor von Kobell for the discrimination of minerals is, almost beyond dispute, the most practical and most reliable that has ever been published. The sixth chapter is nothing but an extract from Prof. v. Kobell's treatise on this subject. It contains all the well-known mineral species, and leads to their determination with almost unerring certainty.

The appended tables, taken from Plattner's work on the Blowpipe, have remained unchanged.

For the material of this compilation, the author is principally indebted to the following works:

C. F. Plattner: The Use of the Blowpipe in the Examination of Minerals, Ores, &c. Translated by J. S. Musprath. 3d ed. London.

J. J. Berzelius: The Use of the Blowpipe in Chemistry and Mineralogy. Translated by J. D. Whitney. Boston.

F. von Kobell: Tafeln zur Bestimmung der Mineralien. 5th ed. München, 1853.

J. D. Dana: A System of Mineralogy. 4th ed. New York, 1854.

John Mitchell: Manual of Practical Assaying. 2d ed. London, 1854.

The author, finally, begs to tender his thanks to his friend, Professor Chandler, of Union College, for the valuable suggestions he has received at his hands, and which he has acted upon to the best of his ability, being fully convinced that by adding the improvements recommended by his friend, the practical utility of this little Manual will be greatly increased.

WILLIAM ELDERHORST.

TROY, N.Y.,

# INTRODUCTION.

In preparing this little Manual, it has been my principal care to adapt it to the use of the beginner. The use of the blow-pipe, though elaborately studied and extensively written on by some of the first chemists and mineralogists of the preceding and the present century, has not yet been duly appreciated. This neglect is, perhaps, owing to the rapid advancement of chemical analysis in the humid way, which furnishes, on the whole, more reliable results, and allows of an easy quantitative determination of the various constituents of a body. But it was overlooked that this mode of analysis absorbs much more time, and requires the use of an extensive set of apparatus, whereas an examination before the blow-pipe is sooner performed; requiring scarcely as many hours as an examination in the humid way requires days, and that with the aid only of a few reagents and instruments of small size. It is for this reason that a knowledge of blow-pipe operations is less valuable for the Chemist by profession than for the Mining-Engineer, the Mineralogist, and the Geologist. A small portable box will hold all the necessary reagents and instruments, so that he may carry them with him on his expeditions and travels, and examine on the spot the minerals which he meets with on his explorations; an advantage which ought, truly, not to be overlooked.

For teachers who have not hitherto devoted much time to instruction in this department, a short exposition of the course which I have followed for a number of years may, perhaps, be desirable. For elementary instruction, the students are only furnished with the principal reagents, viz.: carbonate of soda, salt of phosphorus, borax, and solution of cobalt; of apparatus they want a fluid-lamp, blow-pipe with platinum point, platinum-pointed forceps, platinum wire, charcoal, and closed and open glass tubes. After having explained to them the action of the two cones of the flame, and instructed them in making beads, and conducting the processes of oxidation and reduction, I make them perform the most important operations, and study the behavior of the most commonly occurring substances, with and without fluxes. I give the substances in somewhat the following order:

Sesquioxide of iron, all the reactions given in Table II, 10.
Peroxide of manganese, Table II, 13.
Sesquioxide of chromium, Table II, 6.
Oxide of cobalt, and nickel, Table II, 7, 16.
Protoxide of copper, Table II, 8, and § 37.
Oxide of zinc, Table II, 27, and metallic zinc §§ 25, 45.
Oxide of tin, Table II, 22, and metallic tin § 26.
Oxide of lead, Table II, 12, and metallic lead § 23.
Oxide of bismuth, Table II, 3, and metallic bismuth §§ 17, 22.
Antimonious acid, Table II, 1, and metallic antimony §§ 16, 21.
Arsenous acid, Table II, 2, §§ 9, 15.
Oxide of mercury, Table II, 14.
Alumina, Table I, 5, and § 44.
Magnesia, Table I, 4, and § 44.
Silica, § 39.
A sulphide, §§ 10, 14, 107.
A borate, § 60.
A chloride, §§ 65, 66.

Having performed all these operations, the student will be qualified to enter upon the analysis of substances of not too compound a character. If he meets on his way with bodies, the behavior of which before the blow-pipe he has not previously studied, he will not have any difficulty in determining their character if he follows the directions given in the second chapter. The *modus operandi* will be best understood by a few examples.

1. The substance under examination is sulphide of antimony.

Examination in a matrass: At a very high temperature, a black sublimate is obtained, becoming reddish-brown when cold. In reading over the list in § 10, we find this character belonging to sulphide of antimony.

Examination in an open glass tube: gives sulphurous acid detected by the odor and action on blue litmus-paper, and white fumes which partly condense in the tube. On examining the sublimate with a magnifying-glass, it is found to be amorphous, hence must be antimonious acid (§ 16).

Examination on charcoal alone: is completely volatilized with emission of sulphurous acid, and deposits a white volatile coating, possessing the properties of the coating of antimony (§ 21).

These few operations are quite sufficient to establish the nature of the substance under trial, since the absence of the more fixed metals is proved by the volatility of the substance on charcoal and in the open tube, and the absence of metals giving coatings by the purity of the antimony-coating. The presence of arsenic would have been betrayed by an alliaceous odor when heated on charcoal. The only substance which would have escaped detection by these operations is sulphide of mercury. In order to ascertain its presence or absence, we perform the operation given under "*Mercury*" in Chapter III.

The result giving an answer in the negative, the body was "sulphide of antimony."

2. The substance under examination is chromate of lead.

Examination in a matrass: } fuses and changes color, but
Examination in an open tube: } gives nothing volatile.

Examination on charcoal alone: fuses, gives small metallic globules, and deposits a coating which is lemon-yellow while hot, and sulphur-yellow when cold, indicative of lead (§ 23). It is always desirable to collect the metal to a large globule, and to study its physical properties. This end is best attained by mixing the substance with carbonate of soda and a little borax, and exposing the mixture to the reduction-flame on charcoal. In this particular case, a metallic button is obtained which is soft, may be flattened by the hammer and cut by the knife, properties belonging to metallic lead.

Examination with borax and salt of phosphorus: Before proceeding with this examination it is necessary to test the substance for the presence of sulphur after the method given § 107 (unless the presence of this element was detected by the examination in the open glass tube or on charcoal alone); no sulphur being present, borax and salt of phosphorus beads are made on charcoal, and small portions of the substance added. With both fluxes nearly the same reactions are obtained; in oxydation-flame dark-red while hot, and fine yellowish-green when cold; in reduction-flame green, hot and cold. In order to find out what body produces such reactions, we use Table III, which leads us to sesquioxide of chromium. To corroborate the result, the substance may be fused with carbonate of soda and nitre, as described, § 68.

The physical properties of the body under trial lead to the final conclusion that it must be chromate of lead.

3. The substance is an alloy of silver, copper, and lead.

Examination in a matrass: } no change.
Examination in an open tube: }

Examination on charcoal alone: fuses and deposits a copious coating, which is lemon-yellow while hot and sulphur-yellow when cold, indicative of lead (§ 23); the coating cannot contain any oxide of bismuth, because the color would be darker in this case, but might contain oxide of zinc or oxide of antimony. The test is for the presence of the former, the coating is played upon with the oxydation-flame: it is completely volatile, hence no zinc present (might also be tested with solution of cobalt § 45); to test the coating for the presence of oxide of antimony, it is scraped off from the charcoal and dissolved in a bead of salt of phosphorus, v. § 87, or the alloy is treated with boracic acid as described under the head of "*Antimony*" in Chapter III. If the blast is continued for a long time, a faint dark-red coating is formed near the assay-piece, indicative of silver, § 27, and a dark metallic globule remains.

Examination with borax and salt of phosphorus: the globule remaining on the charcoal after volatilization of the lead, is treated with borax on charcoal in oxidation-flame; the borax becomes colored. Owing to the reducing effect of the charcoal, the influence of the oxidation-flame cannot be well observed on charcoal, hence the borax is removed from the metallic globule, fastened into the hook of a platina wire, and here exposed to the action of the oxidation-flame: the bead is green while hot, and blue when cold. On consulting Table III we find that this reaction is produced by oxide of copper, and by a mixture of oxide of cobalt and sesquioxide of iron: to decide between the two, we now expose the bead to the action of the reduction-flame; it becomes red and opaque, thus proving the presence of oxide of copper.

By the examination on charcoal, *per se*, we were led to suspect the presence of silver; in order to establish this beyond a doubt, we refer to Chapter III, "Silver;" here we find a method (§ 105) by which the presence of silver may be ascertained in compounds of all descriptions. In our case, having to deal only with lead, copper, and silver, the treatment with vitrified boracic acid and metallic lead is, of course, superfluous. We place our alloy at once on the cupel and direct the oxidation-flame upon it; if, after cessation of the rotatory motion, the globule should not possess the bright lustre of silver, some pure metallic lead has to be added, in order to remove the last traces of copper. We finally obtain a bright globule exhibiting all the characteristic properties of silver.

Thus we have established the presence of lead, copper, and silver.

4. The substance under examination is copper nickel, containing arsenic, sulphur, nickel, cobalt, and iron.

Examination in a matrass: gives a slight sublimate, consisting of octahedral crystals, pointing to the presence of arsenic (§ 11).

Examination in a glass tube open at both ends: gives a copious crystalline sublimate of arsenous acid, and a faint odor of sulphurous acid; to establish the presence of sulphur beyond doubt, we refer to Chapter III, "Sulphur," where we find the method (§ 107) for discovering sulphur when in combination with other substances. In performing the test there described, we obtain the sulphur-reaction.

Examination on charcoal alone: gives abundant arsenical fumes, leaving a metallic globule which, even with continued blowing, does not give rise to the formation of a coating on the charcoal (absence of volatile metals).

Having removed all volatile substances, we now proceed to examine the remaining globule. On applying a magnet, we find it powerfully attracted, showing the presence of either iron, nickel, or cobalt, perhaps all of them, either alone or combined with other non-volatile metals. We add some borax to the globule and expose it to the action of the oxidation-flame, then remove the borax from the globule, fasten it into the hook of a platina wire, and here observe the color: green while hot, blue when cold as in the preceding case (example 3), but on exposing the bead to the action of the reduction-flame (which is best done by placing it on charcoal and touching it with tin) it does not become brown and opaque, showing therefore the presence of a small quantity of iron with cobalt. We now add a fresh portion of borax to the metallic globule, in order to see whether it consists entirely of cobalt (that it cannot contain any considerable amount of iron, is proved by the appearance of the cobalt reaction in the first trial, iron being much more readily dissolved by borax than cobalt): the bead is violet while hot, and assumes a brownish color on cooling; by referring to Table III, we see that this effect is produced by nickel containing cobalt. Referring to Chapter III, "Nickel," we find the method to detect the presence of this metal when in combination with iron and cobalt, and also the presence of copper, if the assay should contain a small quantity of it.

By the above examples the use of the methods given in the third chapter will be sufficiently illustrated. If the substance under examination is of a simple composition, its nature is readily ascertained by following the general method laid down in the second chapter; but if the reactions obtained clearly point to the complex nature of the body, we refer to the respective sections of Chapter III; if, for example, we suspect the presence of cobalt in a mineral consisting of arsenides, we test the substance according to § 69; if a small quantity of copper is to be discovered in a mineral, we proceed as directed in § 71, &c.

The student who is willing to devote more time to the subject than is usually allotted to it in our colleges, will do well to go carefully through all the reactions given in the second chapter, and thus familiarize himself with the colors and other properties of the various coatings, sublimates, &c., and also to perform the principal tests by which substances are discovered when in combination with others, which are at length exposed in the third chapter. In order to obtain characteristic reactions, it is important to experiment upon a suitable substance. For the benefit of the beginner, who would naturally be embarrassed in the choice of a body suitable for the experiment, I add a list of substances which, with few exceptions, are readily obtained, and which are sufficient to illustrate all the important reactions. After having mentioned a reaction, or described a process (in Chapter II and III), I have added a number in [ ] brackets. The number points to the substance of the list, below given, best adapted to illustrate the reaction. As each experiment requires only a very small quantity of the substance, they are most conveniently kept in small glass tubes of about an inch and a half in length and one-eighth of an inch in diameter. For the first fourteen substances no glass tubes are required, since they are the regular blow-pipe reagents. A small box containing seventy-five of the little tubes will hold the whole collection.

# COLLECTION OF SUBSTANCES,

*Well adapted to illustrate the important Reactions of Bodies before the Blowpipe.*

1. Carbonate of soda.
2. Borax.
3. Salt of phosphorus.
4. Bisulphate of potassa.
5. Boracic acid.
6. Fluor spar.
7. Nitrate of cobalt.
8. Oxalate of nickel.
9. Oxide of copper.
10. Chloride of silver.
11. Lead.
12. Iron.
13. Tin.
14. Bone-ash.
15. Chloride of potassium.
16. Bromide of potassium.
17. Iodide of potassium.
18. Chloride of sodium.
19. Chloride of ammonium.
20. Chlorate of potassa.
21. Alumina.
22. Sulphate of copper.
23. Nitrate of lead.
24. Oxide of antimony.
25. Arsenous acid.
26. Oxide of bismuth.
27. Oxide of cadmium.
28. Sesquioxide of chromium.
29. Oxide of cobalt.
30. Protoxide of mercury.
31. Molybdic acid.
32. Oxide of silver.
33. Binoxide of tin.
34. Tungstic acid.
35. Sesquioxide of uranium.
36. Oxide of zinc.
37. Chloride of copper.
38. Arsenite of copper.
39. Subchloride of mercury.
40. Protochloride of mercury.
41. Antimony.
42. Arsenic.
43. Bismuth.
44. Cadmium.
45. Silver.
46. Zinc.
47. Alloy of mercury and tin.
48. Alloy of lead and antimony.
49. Alloy of lead and bismuth.
50. Alloy of lead and zinc.

51. Alloy of lead, copper, and silver.
52. Alloy of tin and copper.
53. Alloy of zinc and cadmium.
54. Rock crystal.
55. Gypsum.
56. Calc-spar.
57. Strontianite.
58. Whitherite.
59. Magnesite.
60. Mica.
61. Felspar.
62. Albite.
63. Petalite.
64. Hematite.
65. Rutile.
66. Pyrolusite.
67. Lepidolite.
68. Apatite.
69. Franklinite.
70. Pitchblende.
71. Chromic iron.
72. Cerusite.
73. Malachite.
74. Gray antimony.
75. Iron pyrites.
76. Copper pyrites.
77. Mispickel.
78. Smaltine.
79. Cobaltine.
80. Realgar.
81. Cinnabar.
82. Copper nickel.
83. Molybdenite.
84. Berthierite.
85. Bournonite.
86. Tetrahedrite.
87. Onofrite, or Clausthalite.
88. Sulphides of arsenic and antimony (artificial).
89. Sulphides of arsenic, antimony, lead, and copper (artificial).

# TABLE OF CONTENTS.



## FIRST CHAPTER.



## SECOND CHAPTER.

## THIRD CHAPTER.

## FOURTH CHAPTER.



## FIFTH CHAPTER.

## SIXTH CHAPTER.



## TABLES.



---

## ABBREVIATIONS.

OFl for Oxydation-flame; RFl for Reduction-flame; SPh for Salt of Phosphorus; Bx for Borax; Sd for Carbonate of Soda; SoCo or SCo for Solution of Nitrate of Cobalt; Ch for Charcoal; Ct for Coating; Blp for Blow-pipe; H for Hardness; G for Specific Gravity

# A MANUAL

OF

# BLOW-PIPE ANALYSIS.

---

## FIRST CHAPTER.

### AUXILIARY APPARATUS AND REAGENTS.

§ 1. THE common blow-pipe of gas-fitters, jewellers, &c., is not very well adapted for analytical researches, as the narrow outlet becomes frequently obstructed by the moisture which is exhaled from the lungs and condenses in the tube. To avoid this inconvenience, the long cylindrical tube of the blow-pipe should be furnished at the extremity with a globular or cylindrical chamber for the reception of the condensed water. In this chamber the jet is inserted at a right angle to the tube. Silver is, in many respects, the best material for the construction of a blow-pipe, but has the disadvantage of becoming very hot when used for a long while, so that it becomes almost impossible to hold it with the naked fingers; next to silver stands German silver and brass. For jets, platinum is preferable to all other metals. A mouth-piece of box-wood or ivory is convenient, though not necessary.

§ 2. Any kind of flame may be used for the blow-pipe, provided it be not too small. Some of the older chemists used common candles in preference, and it must be confessed that, in the majority of cases, the heat produced by the flame of a good sperm candle is quite sufficient. Berzelius recommended an oil lamp with a flat wick, which is now in general use as "Berzelius's Blow-pipe Lamp." I find that a common fluid lamp, with a rather large burner, answers every purpose; it gives a very good heat, and, besides being much cleaner than an oil lamp, admits of a very quick and accurate adjustment of the size of the flame, by means of a little brass cylinder, which is movable, and slides up and down the burner. The heating of substances in glass tubes and matrasses is best performed over a common spirit lamp.

§ 3. As supports, charcoal, platinum, and glass are principally used. Wood charcoal is in most cases the best support. It must be well burnt, and not scintillate or smoke; it must leave but little ash; charcoal of light wood, as alder, &c., has been found the best.

Platinum is used whenever the reducing action of the charcoal acts injuriously. It is advantageously employed on all occasions where no reduction to the metallic state takes place, since the color of the flux is much better seen on the platinum than on charcoal. It is mostly used in the shape of wire, the end of which is bent so as to form a hook, which serves as support to the flux. As foil, its use is very limited. A little platinum spoon, of from about 12 to 15 m. m. in diameter, is very convenient for fusing substances with bisulphate of potassa or nitre.

Glass tubes, open at both ends, are used for calcination, and for testing the presence of substances which are volatile at a high temperature. The tubes should be from 5 to 8 inches long. Of glass tubes, sealed at one end, or little matrasses, an assortment should always be kept on hand, since they are of very frequent use.

§ 4. Of other apparatus, the most necessary are:

A mortar of agate or chalcedony, from 1½ to 2 inches in width, with pestle of the same material.

A forceps of brass or German silver, with platinum points.

A forceps of steel.

A little hammer and anvil, both of steel and well polished.

A three-cornered file for cutting glass tubes, trying the hardness of minerals, &c.

A little magnet.

A pocket magnifying glass.

A set of watch glasses, which are very convenient for the reception of the assay-piece, the metallic globules, &c.

§ 5. Of reagents, Carbonate of Soda, Borax, and Salt of Phosphorus, are the most important ones; but there are others, which, though not so extensively used, still are indispensable for the detection of certain substances; those only shall here be mentioned; others, the use of which is very limited, are omitted in this list.

Carbonate of Soda: The monocarbonate or the bicarbonate may be indifferently employed; it must be perfectly free from sulphuric acid, for the presence of which it may be tested as shown § 107. The neutral oxalate of potassa and the commercial [fused] cyanide of potassium deserve in many cases the preference, their reducing powers being superior to that of carbonate of soda.

Borax: The commercial article is purified by crystallization, the crystals dried and reduced to a coarse powder.

Salt of Phosphorus: [double phosphate of soda and ammonia] 100 parts of crystallized common phosphate of soda and 16 parts of sal ammoniac are dissolved in 32 parts of water; the solution is aided by heat, the liquid filtered while hot, and the crystals, which form on cooling, dried between blotting paper. When pure it gives a glass which, on cooling, remains transparent; if this is not the case it must be purified by recrystallization. It is kept as a coarse powder.

Bisulphate of Potassa: It is employed in the fused [anhydrous] state as a coarse powder; it must be kept in a bottle provided with a ground glass stopper.

Vitrified Boracic Acid: Is employed in the state of a coarse powder.

Fluor-spar: Must be deprived of water by ignition; must be perfectly free from boracic acid, which may be tested as described § 61. It is convenient, to keep in a separate bottle a mixture of 1 part of finely powdered fluor-spar with 4½ parts of bisulphate of potassa.

Nitrate of Cobalt, in solution: It must be pure, free from alkali, and [for many purposes] free from nickel; it is kept in a bottle with a ground glass stopper, which, very conveniently, is so much elongated as to dip into the liquid. Instead of the nitrate, the oxalate of cobalt may be used, which, being in the shape of powder, is advantageously substituted for the former in travelling.

Nitrate or Oxalate of Nickel: It must be perfectly free from cobalt; it is tested with borax, with which it ought to produce a pure brown glass.

Oxide of Copper: It is best prepared by heating the nitrate to ignition.

Chloride of Silver: It is prepared by precipitating a solution of nitrate of silver with hydrochloric acid, washing the precipitate, and making it into a thick paste with water, which is kept in a small glass-stoppered bottle. This reagent should not be used with platina wire, since the silver fuses with the platina to an alloy; thin iron wire is in this case substituted for the platina. For each experiment a fresh hook should be made.

Pure Metallic Lead: It is easily obtained pure by decomposing a solution of the acetate by metallic zinc; the precipitate is repeatedly washed with water and then dried between blotting paper.

Metallic Iron: In the shape of thin wire [harpsichord wire].

Metallic Tin: Usually in the shape of foil, which is cut into strips and rolled up tightly.

Bone-ash: In the state of very fine powder.

Test Paper: Blue and red Litmus Paper, and Brazil Wood Paper.

§ 6. If the analytical research is strictly confined to blow-pipe operations, the above enumerated reagents are sufficient; but if, as is sometimes advantageously done, some simple operations of the humid method of chemical analysis are called to aid, the list must be somewhat extended. The most important of these reagents, all of which must be kept in bottles with ground glass stoppers, are: Sulphuric Acid, Hydrochloric Acid, Nitric Acid, Oxalic Acid, Hydrate of Potassa, Ammonia, Carbonate of Ammonia, Chloride of Ammonium, Molybdate of Ammonia, Ferrocyanide of Potassium, Ferridcyanide of Potassium, Bichloride of Platinum, Acetate of Lead, Sulphuretted Hydrogen Water, Sulphydrate of Ammonia, Alcohol, Distilled Water.

The principal auxiliary apparatus are: Test-tubes and Test-tube Rack, small Porcelain Dishes, small Beaker Glasses, some Glass Funnels and Filtering Stand, Filtering Paper, Platinum Crucible, some Glass Rods and round Glass Plates, for covering beaker glasses, a common Spirit Lamp, and a Spirit Lamp with Argand Burner.

# SECOND CHAPTER.

## GENERAL ROUTINE OF BLOW-PIPE ANALYSIS.

§ 7. On examining a substance before the blow-pipe, with a view to determine its nature or to ascertain the presence or absence of certain matter, it is advisable to follow a systematic way, composed of a series of operations, and to attentively observe the changes which the body undergoes under the influence of the various agents which are brought to act upon it. The various operations which the assay is submitted to are so many questions, to which the phenomena which we observe constitute so many answers; and from their appearance or non-appearance, we are able to draw definite conclusions as to the nature of the substance under examination.

The following order, and the rules to be observed in the execution of the various operations are, essentially, the same as first pointed out and laid down by Berzelius.

1. Examination in a glass tube, sealed at one end, or a matrass.
2. Examination in a glass tube open at both ends.
3. Examination on charcoal *per se*.
4. Examination in the platinum-pointed pincers, or on platinum wire *per se*.
5. Examination with borax, and salt of phosphorus.
6. Examination with carbonate of soda.
7. Examination with solution of cobalt.

Regarding the size of the assay, a piece the size of a mustard seed will generally be found sufficient; larger pieces, without showing the reaction more distinctly, requiring so much more labor. In some cases, however, it is advantageous to employ a greater quantity, ex. gr. in reductions, or in heating in a glass tube; for the larger the metallic globule, and the greater the amount of the sublimate produced, the more readily can its nature be ascertained.

*Examination in a closed Glass Tube, or a Matrass.*

§ 8. The assay-piece is introduced into a glass tube, sealed at one end, or into a small matrass, and heat applied by means of a common spirit lamp. The heat must at first be very low, but may be gradually raised to redness, if necessary. By this treatment we learn:

§ 9. 1. Whether the substance is entirely or partly volatile or not. Among the phenomena to be observed, the following are deserving of particular attention:

The substance gives out water, which partly escapes and partly condenses in the colder portion of the tube. This points to the presence of a salt containing water of crystallization [No. 22], or to the presence of a hydrate, or to such salts which contain water mechanically inclosed between the laminæ of the crystals [No. 18]; in this case the body usually decrepitates. The drops of condensed water are to be examined with test-paper: an alkaline reaction denotes the presence of ammonia, an acid reaction the presence of some volatile acid, as sulphuric, nitric, hydrochloric, hydrofluoric acid, &c.

§ 10. The substance gives out a gas or vapor. Those of most usual occurrence are:

*a.* Oxygen, easily recognized by rekindling into flame a match which has been extinguished, so as to leave only a spark at the extremity; points to the presence of a peroxide, nitrate, chlorate, bromate or iodate [No. 20].

*b.* Sulphurous acid, easily recognized by its peculiar odor and action on blue litmus paper; indicates the presence of a sulphate or sulphite [No. 22].

*c.* Sulphuretted hydrogen, recognized by its peculiar odor; indicates the presence of sulphides containing water.

*d.* Nitrous acid or peroxide of nitrogen, recognized by its deep orange red color and acid reaction; indicates the presence of a nitrite or nitrate [No. 23].

*e.* Carbonic acid, recognized by causing a turbidity in a drop of lime-water suspended from a watch-crystal and exposed to the escaping gas; points to the presence of a carbonate.

*f*. Cyanogen, recognized by its peculiar odor and by burning with a crimson flame; indicates the presence of a cyanogen-compound.

*g*. Ammonia, recognized by its odor and alkaline reaction; indicates the presence of an ammoniacal salt or of an organic nitrogenous substance; in the latter case, the mass usually blackens, and evolves at the same time either cyanogen or empyreumatic oils of offensive odor [No. 3].

§ 11. The substance yields a sublimate. The sublimate is either white or possessed of a peculiar color. White sublimates are formed by

*a*. Many salts of ammonia; on removing the sublimate from the tube, placing it on a watch-crystal, adding a drop of hydrate of potassa, and applying heat, ammonia is evolved [No. 19].

*b*. The chlorides of mercury; the subchloride sublimes without previous fusion; the protochloride fuses first, then sublimes; the sublimate is yellow while hot, but becomes white on cooling [Nos. 39 and 40].

*c*. Oxide of antimony; it fuses first to a yellow liquid, then sublimes; the sublimate consists of lustrous needle-shaped crystals [No. 24].

*d*. Arsenous acid; the sublimate consists of octahedral crystals [No. 25].

*e*. Tellurous acid; shows a reaction similar to that of oxide of antimony, but requires a much higher temperature; the sublimate is amorphous.

Sublimates possessed of metallic lustre, so-called metallic mirrors, are formed by:

*a*. Metallic arsenic and arsenides containing more than one equivalent of arsenic to two of metal, also, some sulph-arsenides [No. 77]; cutting the tube below the sublimate, and exposing the mirror to gentle heat in the flame of a spirit lamp, the peculiar odor of arsen c is perceived.

*b* ercury, amalgams, and some salts of mercury; the sublimate consists of minute globules of metallic mercury, which, by friction with a piece of copper wire, readily unite to larger globules [No. 47].

*c*. Some alloys of cadmium.

*d*. Tellurium; only at a very high temperature; the sublimate consists of small globules, which solidify on cooling.

Sublimates possessed of distinct color are formed by:

*a.* Sulphur and sulphides containing a large amount of sulphur; the sublimate is from deep-yellow to brownish-red while hot, but pure sulphur-yellow when cold [No. 75].

*b.* The sulphides of antimony, alone or in combination with other sulphides; the sublimate forms only at a very high temperature, and is deposited at a short distance from the assay-piece; it is black while hot, reddish-brown when cold [No. 74].

*c.* The sulphides of arsenic and some compounds of metallic sulphides with arsenides; the sublimate is dark brownish-red while hot, but from reddish-yellow to red when cold [No. 80].

*d.* Cinnabar; the sublimate is black, without lustre, and yields a red powder on being scratched with a knife [No. 81].

*e.* Selenium and some selenides; the sublimate appears only at a high temperature, is of a reddish or black color, and yields a dark-red powder; at the open end of the tube the peculiar odor of selenium (resembling rotten horse-radish) is perceived [No. 87].

§ 12. 2. Whether the substance undergoes any change, or remains unaltered.

Many substances, under this treatment, suffer physical changes without being affected in their chemical constitution. A great many minerals, when heated, decrepitate; others phosphoresce, as fluor-spar and apatite. The most important of these physical changes, is that of color: from white to yellow, and white again on cooling, points to oxide of zinc [No. 36]; from white to yellowish-brown, dirty pale-yellow on cooling, points to oxide of tin [No. 33]; from white to brownish-red, yellow when cold, and fusible at a red heat, points to oxide of lead [No. 72]; from white to orange-yellow or reddish-brown, pale yellow when cold, and fusible at a bright red heat, points to teroxide of bismuth; from red to black, and red again on cooling, points to sesquioxide of iron [No. 64].

### *Examination in a Glass Tube open at both ends.*

§ 13. The assay-piece is introduced into the tube to a depth of about half an inch, the end to which it lies nearest slightly inclined, and heat applied. The air contained in the tube becomes heated;

it rises, escapes from the upper end, and fresh air enters from below. In this manner a calcination is effected, and many substances which remained unchanged when heated in a matrass, yield sublimates or gaseous products when subjected to this treatment, owing to the formation of volatile oxides.

By this means we can easily detect the presence of the following substances:

§ 14. Sulphur; sulphurous acid is formed, which is characterized by its peculiar odor and action on moistened litmus paper [No. 75].

§ 15. Arsenic; if present in sufficient quantity it yields a white and very volatile sublimate of arsenous acid, consisting of minute octahedral crystals; by application of gentle heat it may be driven from one place to another [No. 77].

§ 16. Antimony; white fumes of antimonious acid are given out which partly escape, and partly condense in the upper part of the tube. The sublimate is a white powder, and may, if consisting of pure antimonious acid, be volatilized by heat. In most cases, however, the oxidation proceeds farther, and the antimonate of the oxide of antimony, a non-volatile white powder, is formed [No. 41].

§ 17. Metallic Bismuth; it is converted into oxide, which condenses at a short distance from the assay, and which, by heat, may be fused to brownish globules [No. 43].

Mercury and amalgams; yield sublimates of metallic mercury [No. 47].

§ 18. Tellurium and tellurides; tellurous acid is produced, which condenses in the upper part of the tube to a white non-volatile powder; on application of heat it fuses to colorless globules.

Selenium and selenides; evolve a gaseous oxide of a peculiar odor, resembling that of rotten horse-radish [No. 87].

### *Examination on Charcoal* per se.

§ 19. In examinations of this kind, particular attention must be paid to the odor of the escaping gases, and to the color and other properties of the rings, or coatings, which form on the charcoal around the assay-piece. The interior [reduction flame, R Fl] and exterior [oxidation flame, O Fl] cones of the flame acting in an opposite sense, the phenomena produced will be very different; hence, two assays should always be made, exposing the substance first to the action of the O Fl and then to the action of the R Fl. The following bodies undergo, when submitted to this treatment, characteristic changes.

§ 20. Arsenic: It is volatilized without previous fusion, the Ch is covered with a white Ct, which is far distant from the assay-piece, and which is produced by both the O Fl and R Fl; the Ct is very volatile, and is easily driven away by the Blp flame, emitting the peculiar alliaceous odor characteristic of arsenic [No. 42].

§ 21. Antimony: It enters readily into fusion and covers the Ch with white oxide; the ring is not so far distant from the assay-piece as in the case of arsenic; it may be driven away by the Blp flame, but is not so volatile as that of arsenic, and does not emit an alliaceous odor. Metallic antimony, when fused on Ch and heated to redness, remains a considerable time in a state of ignition without the aid of the Blp, disengaging, at the same time, a thick white smoke, which is partly deposited on the Ch around the metallic globule in white crystals of a pearly lustre [No. 41].

§ 22. Bismuth: It fuses readily in both flames and covers the Ch with oxide, which is dark orange-yellow while hot and lemon-yellow when cold. The yellow Ct is usually surrounded by a white ring, consisting of carbonate of bismuth. The Ct is somewhat nearer the assay than that of antimony; it may be driven away by both flames; but the oxide of antimony, when played upon with the R Fl, imparts to the flame a greenish-blue tinge, which the oxide of bismuth does not [No. 43].

§ 23. Lead: It fuses easily and coats the Ch in both flames with oxide, which is dark lemon-yellow while hot and sulphur-yellow when cold; in thin layers it is bluish-white and consists of carbonate. The Ct is found at the same distance from the assay as that of bismuth; it may be driven away by both flames; when played upon with the R Fl it imparts to the flame an azure-blue color [No. 11].

§ 24. Cadmium: It fuses readily; exposed to the O Fl it burns with a dark-yellow flame, emitting brown fumes of oxide which cover the Ch around the assay. This Ct is very characteristic; it is, when cold, of a reddish-brown color, in thin layers orange-yellow;

it is easily volatilized by both flames, without imparting a color to them [No. 44].

§ 25. Zinc: It fuses readily; exposed to the O Fl it burns with an intensely luminous greenish-white flame, emitting at the same time a thick white smoke which, partly condensing on the Ch, covers it with oxide, yellow while hot and white when cold. The Ct when played upon with the O Fl becomes luminous, but does not disappear [No. 46].

§ 26. Tin: It fuses readily; exposed to the O Fl it is converted into oxide, which may be blown away and thus be made to appear as a Ct; it is always found closely surrounding the assay-piece, is slightly yellow and luminous while hot, white when cold, and non-volatile in both flames. Exposed to the R Fl the molten metal retains its bright metallic aspect [No. 13].

§ 27. Silver: When exposed for a long time to the action of the R Fl it yields a slight dark-red Ct of oxide [No. 45].

§ 28. Selenium: It fuses very readily in both flames with disengagement of brown fumes; at a short distance from the assay a steel-gray Ct of a feeble metallic lustre is deposited; played upon with the R Fl it disappears with emission of a strong odor of rotten horse-radish, at the same time imparting to the flame a fine blue color [No. 87].

§ 29. Tellurium: It fuses very readily and coats the Ch in both flames with tellurous acid; the Ct is not very far distant from the assay; it is of a white color with a red or dark-yellow edge; played upon with the R Fl it disappears, imparting to the flame a green tinge.

§ 30. Besides the above named metals there are some other substances which, when treated before the Blp upon Ch cover it with coatings, which may be driven away when played upon with the O Fl, and which show in many cases a great resemblance to the Ct produced by antimony. Among the bodies possessing this property the following are the most frequently occurring ones:

The sulphurides of potassium, sodium, and lithium.

The chlorides of ammonium, potassium, sodium, and lithium.

The chlorides of mercury, antimony, zinc, cadmium, lead, bismuth, tin, and copper.

The bromides and iodides of potassium and sodium.

*Examination in the Platinum-pointed Pincers.*

§ 31. This experiment serves a double purpose. It acquaints us with the degree of fusibility of the assay, and shows the presence or absence of such substances which possess the property of imparting to the flame a peculiar color. Many metals, the sulphurides, and some other compounds act upon metallic platinum at a high temperature; the fusibility, &c., of such substances ought to be tested on Ch. Others, again, fuse so easily that they cannot be held a sufficiently long time between the pincers to observe the color which they impart to the flame; they are most conveniently attached to the hook of the platinum wire, which is best done by heating the wire to redness and then touching the powder of the assay with it; a sufficient quantity generally remains adhering to the wire.

Some minerals decrepitate violently as soon as they are touched with the flame; in such cases, Berzelius advises to powder the substance very finely in an agate-mortar with addition of a little water, to place one or two drops of the mixture on a piece of Ch, and to gently heat it by means of the Blp flame until the mass lies loosely upon the Ch; it may then be taken up and held by the pincers. The same process is advantageously employed with substances which fuse only at a very high temperature. In all other cases the substance is roughly powdered and a thin piece which shows prominent edges selected for the experiment.

The assay is exposed to the action of the inner cone of the flame, when the outer cone may exhibit the following changes of color:

§ 32. 1. Yellow.

Soda and its salts cause an enlargement of the outer flame, and impart, at the same time, an intense reddish-yellow color [No. 18]. The presence of other substances which also possess the property of coloring the flame, but not in so high a degree, does not prevent the reaction. Silicates containing soda, exhibit the same phenomenon to a smaller or greater extent, according to their degree of fusibility and the amount of soda which they contain [No. 62]. With many salts of soda, which do not exhibit the reaction very distinctly, it can be produced by mixing the salt with some chloride of silver to a paste (v. § 5), fastening it to the hook of a thin iron-wire, and then exposing it to the action of the inner flame.

§ 33 2. Violet.

Potassa and many of its salts impart to the outer flame a distinct violet color [No. 15]. The presence of a small quantity of a salt of soda or lithia prevents the reaction. An addition of chloride of silver favors the reaction with the carbonate, nitrate, and some other salts of potassa.

§ 34. 3. Red.

Lithia and its salts impart to the outer flame a fine carmine-red color [No. 63]; the chloride of lithium shows the reaction better than any other salt. The presence of a salt of potassa does not prevent the reaction; the presence of even a small quantity of a salt of soda changes the color to yellowish-red. An addition of chloride of silver favors the reaction with many salts of lithia.

Chloride of strontium and some other salts of strontia, ex. gr. the carbonate and the sulphate, color the outer flame, immediately or after a while, carmine-red [No. 57]. The presence of baryta prevents the reaction. The carbonate and sulphate of strontia show the reaction remarkably well when mixed with chloride of silver and heated on iron-wire (v. § 5).

Chloride of calcium, calcareous spar, many compact limestones, and fluor-spar, color the outer flame, immediately or after a while, red; the color is not so intense as that produced by strontia. Gypsum and anhydrite impart at first a pale yellow, afterwards a red color of little intensity [No. 56]. An addition of chloride of silver usually increases the intensity of the color.

§ 35. 4. Green.

Chloride of barium, carbonate and sulphate of baryta, color the outer flame yellowish-green. The presence of lime does not prevent the reaction [No. 58]. An addition of chloride of silver makes the color much more intense.

Oxide of copper and some of its salts, ex. gr. the carbonate, sulphate, and nitrate, impart to the outer flame a fine emerald-green color. Iodide of copper and some silicates containing copper, ex. gr. dioptase and chrysocolla, act in the same manner [No. 73]

An addition of chloride of silver produces increased intensity of color.

Phosphoric acid, phosphates, and minerals containing phosphoric acid, impart to the outer flame a bluish-green color [No. 3].

Boracic acid colors the outer flame yellowish-green (greenfinch color) [No. 5]; if a small quantity of soda is present the color is mixed with yellow.

Molybdic acid, oxide of molybdenum, and the native sulphide of molybdenum, color the outer flame yellowish-green, like baryta [No. 83].

Tellurous acid enters into fusion, emits white fumes, and colors the outer flame green.

§ 36. 5. Blue.

Arsenic and some arsenides, ex. gr. smaltine and copper-nickel [No. 82] when heated on Ch impart a light-blue color to the outer flame. Some arsenates, ex. gr. scorodite and cobalt bloom, exhibit the same phenomenon in the pincers.

Antimony, fused on Ch in R Fl is surrounded by a very feeble bluish light [No. 41].

Metallic lead, fused on Ch in R Fl is surrounded by an azure-blue light. Many salts of lead, heated in the pincers or on platinum wire, impart an intense azure-blue color to the outer flame [No. 11].

Chloride of copper colors the outer flame intensely azure-blue; after a while the color becomes green, owing to the formation of oxide of copper [No. 37].

Bromide of copper colors the outer flame greenish-blue; after a while the color changes to green.

Selenium, fused on Ch in R Fl vaporizes with an azure-blue light.

## *Examination with Borax and Salt of Phosphorus.*

§ 37. The examination of the assay with borax and salt of phosphorus is eminently adapted to detect the presence of metallic oxides, a great number of them possessing the property of being at a high temperature dissolved by these fluxes with a peculiar color. Unoxidized metals and metallic sulphides, arsenides, &c., differ in this respect very materially from the pure oxides; hence it is necessary, before performing the experiment, to convert all such substances into oxides. This is effected by calcination, on Ch or in an open glass tube. The finely powdered assay is placed on Ch and alternately treated with the O Fl and R Fl, and this process is repeated until the substance no longer emits, while in the incandescent state, the odor of sulphur or arsenic. The heat must never be raised so high as to cause fusion, and between every two succeeding calcinations the assay should be taken from the Ch and freshly powdered.

The experiment is generally made on platinum wire, where the color of the bead is more readily observed; Ch is used only in such cases where the substance under examination contains metallic oxides which are easily reduced. It is not sufficient to observe the color of the bead after cooling; but all changes of color which take place during the action of the flame, and through all the various stages of cooling, should be carefully noticed.

Some substances possess the property of forming a limpid glass with borax, which preserves its transparency on cooling, but which, if slightly heated in the O Fl, becomes opaque, when the flame strikes it in an unequal or intermittent manner. This operation has received the name of "flaming," and any substance thus acted upon is said to become "opaque by flaming."

The third and fourth columns of Tables I and II exhibit the behavior of the most important oxides to borax and salt of phosphorus.

In Table III the oxides are arranged with reference to the color which they impart to the beads in O Fl and R Fl.

### *Examination with Carbonate of Soda.*

§ 38. In subjecting a body to the treatment of Sd we have to direct our attention to two points.

Some substances unite with Sd to fusible compounds, others form infusible compounds, and others again are not acted upon at all; in the last case the Sd is absorbed by the Ch and the assay is left unchanged. With Sd unite to fusible compounds with effervescence:

§ 39. Silicic acid; it fuses to a transparent glassy bead which, after cooling, remains transparent if the Sd has not been added in excess [No. 54].

Titanic acid; it fuses to a transparent glassy bead which, when cold, is opaque and of crystalline structure [No. 65].

Tungstic and molybdic acids; the mass, after the union has been effected, is absorbed by the Ch [No. 31 and No. 34].

The salts of baryta and strontia form with Sd fusible compounds which are absorbed by the Ch [No. 57 and No. 58].

§ 40. The second point to be observed is the elimination of metallic matter. Of the metallic oxides, when treated with Sd on Ch in R Fl, are reduced: the oxides of the noble metals and the oxides of arsenic, antimony, bismuth, cadmium, copper, cobalt, iron, lead, mercury, nickel, tin, zinc, molybdenum, tungsten, and tellurium. Of these, arsenic and mercury vaporize so rapidly that frequently not even a coating is left on the Ch. Antimony, bismuth, cadmium, lead, zinc, and tellurium are partly volatilized and form distinct coatings on the Ch. The non-volatile reduced metals are found mixed up with the Sd. To separate them from the adhering S and Ch powder, we may proceed in the following manner:

The fused mass of Sd and metal, and the portion of the Ch in, mediately below and around the assay, is placed in the little agate mortar, rubbed to powder, the powder mixed with a little water, and stirred up. The heavy metallic particles settle to the bottom, part of the Sd dissolves, and the Ch powder remains suspended in the water. The liquid is carefully poured off, and the residue treated repeatedly in the same manner until all foreign matter is removed. The metal remains behind as a dark heavy powder or, when the metal is ductile and easily fusible, in the shape of small flattened scales of metallic lustre. If the substance under examination contains several metallic oxides, the metallic mass obtained is usually an alloy, in which the several metals may be recognized by processes to be described hereafter. It is only in some exceptional cases that separate metallic globules are obtained, ex. gr. in substances containing iron and copper.

For a more detailed account of the behavior of the various metallic oxides under this treatment, see the second column of Tables I and II.

§ 41. The examination with Sd is usually performed on Ch in the R Fl, and, as a general rule, the flux is added successively in small portions. This is particularly necessary when the assay is to be tested for its fusibility with Sd, since a great many minerals, &c., behave very differently with different quantities of the flux.

§ 42. Instead of carbonate of soda, the neutral oxalate of potassa or cyanide of potassium may be advantageously used for all experiments of reduction, since these reagents exercise a more powerful reducing action than common Sd. They are, for this reason, frequently employed when the presence of such metallic oxides is suspected, whose conversion into metals require high temperatures and the aid of a very efficient deoxidizing agent.

### *Examination with Solution of Cobalt.*

§ 43. A few substances, when moistened with a solution of nitrate of cobalt and exposed to the action of the O Fl, assume a peculiar color. The use of this test is, however, very limited, since the reaction can only clearly be seen in such bodies which, after having been acted upon by the O Fl, present a white appearance, or nearly so.

§ 44. Substances which are sufficiently porous to imbue a liquid, are merely moistened with a drop of S Co, placed into the platinum-pointed pincers, and treated with the O Fl. Other substances must be powdered, the powder placed on Ch, wetted with a drop of S Co, and treated as above. The color can only be distinguished after cooling. A bluish color, of more or less purity, indicates the presence of alumina [No. 21]; and a pale-reddish color [flesh-color] that of magnesia [No. 59]. It must, however, be borne in mind, that the alkaline and some other silicates, when heated with S Co to a temperature above their fusing point, also assume a blue color, owing to the formation of silicate of cobalt. In testing for alumina, therefore, the heat must not be raised so high as to cause fusion of the assay. In testing for magnesia this precaution is not necessary; on the contrary, the color will appear the brighter and the more distinct, the higher the temperature to which the assay was exposed.

§ 45. Among the oxides of the heavy metals, those of zinc and tin assume characteristic colors with S Co. The reaction is best seen when the assay, alone or mixed with Sd, is exposed to the R Fl on Ch. The ring of oxide which is deposited around the assay is then moistened with S Co and treated with the O Fl. Oxide of zinc takes a fine yellowish-green, and oxide of tin a bluish-green color [No. 36 & No. 33].

§ 46. Besides the compounds above mentioned there are some others which, when exposed to the action of S Co and O Fl, experience a change of color. These bodies are either of very rare occurrence, or the change produced in them is not sufficiently distinct. It will, therefore, be sufficient merely to mention the names of the compounds and the color which S Co imparts to them:

Baryta [brownish-red], tantalic acid [flesh-color], zirconia and phosphate of magnesia [violet], titanic acid, niobic acid, and antimonic acid [green], strontia, lime, glucina, and pelopic acid [gray].

# THIRD CHAPTER.

## SPECIAL REACTIONS FOR THE DETECTION OF CERTAIN SUBSTANCES WHEN IN COMBINATION WITH OTHERS.

§ 47. THE preceding chapter and accompanying table show the changes which many of the simple chemical compounds undergo when heated, or when treated with the usual blow-pipe reagents. The reactions are sufficiently characteristic to distinguish the various compounds from each other, so that, when any of the above named substances in a pure state is under examination, there is no difficulty to determine its nature. This, however, is not of frequent occurrence, and in the majority of cases the body to be tested will be of a more complex nature. The results of the experiments will vary accordingly. For instance, an ore of cobalt, containing iron, will not impart to the bead of Bx or S Ph in the O Fl a blue color, but a green one, resulting from the mixture of the blue of cobalt and the yellow of iron; lead, when accompanied by antimony, deposits a dark-yellow coating on Ch resembling that of bismuth, &c. In such cases we may often, by attentively observing all the phenomena which present themselves, and by carefully comparing the results obtained by the various experiments, detect many, if not all, of the components of the substance under examination. Sometimes we attain this end quicker by varying the order, or by introducing auxiliary agents into the series of experiments; and in other cases, again, it is only to be arrived at by subjecting the assay to treatments different from those mentioned in the preceding pages.

This chapter contains the principal reactions for the detection of substances which require the application of peculiar agents, and the methods for ascertaining the presence of certain bodies when in combination with others. The alphabetical arrangement will be found of practical use.

*Ammonia.*

§ 48. Small quantities of ammonia are best detected by mixing the powdered assay [No. 19] with some carbonate of soda or caustic potassa, introducing the mixture into a glass tube, sealed at one end, and applying heat. The escaping gas is characterized by its odor, and by its action on reddened litmus paper. From the appearance of this reaction we are, however, not authorized to infer the preëxistence of ammonia in the assay, since from organic matter containing nitrogen, when subjected to this treatment, ammonia is evolved as a product of decomposition.

*Antimony.*

The reactions of antimony and its compounds, see § 11, § 16, § 21, § 36, and Table II, 1.

§ 49. In presence of lead or bismuth, antimony can not be detected by its Ct on Ch. In this case the metallic compound [No. 48, or No. 85] is treated with vitrified boracic acid on Ch, the flame being so directed that the glass is always kept covered with the blue cone, the metallic globule being on the side; by this means the metals become oxidized, the oxides of lead and bismuth are absorbed by the boracic acid, and the antimonious acid will form a ring on the Ch, provided the temperature was not raised too high.

§ 50. When combined with metals from which it is not easily separated, ex. gr. copper, the evaporation of the antimony takes place so slowly that no distinct Ct is produced. In this case the assay [No. 86] is treated with S Ph on Ch in the O Fl, until the antimony, or at least part of it, has become oxidized and entered into the flux. The glass is now removed from the metallic globule and treated on another place of the Ch with metallic tin in the R Fl; the presence of antimony will cause the glass to turn gray or black on cooling [Table II, 1]. Bismuth behaving under these circumstances in the same manner, the presence of this metal makes the reaction not decisive for antimony. The humid way has then to be resorted to.

§ 51. When the oxides of antimony are accompanied by such metallic oxides which, when reduced on Ch, fuse with the metallic antimony to an alloy, as is ex. gr. the case with the oxides of tin and copper, the latter cannot be recognized by a simple reduction. The oxides have to be treated with a mixture of Sd and Bx on Ch in the R Fl. The little metallic globules are separated from the flux, and fused with from three to five times their own volume of pure lead and some vitrified boracic acid in the R Fl, care being taken to play with the flame only on the glass. Antimonious acid is volatilized, depositing the characteristic ring, while the oxides of the other metals are absorbed by the boracic acid.

§ 52. The sulphides of antimony, when heated in the open glass tube, show the reaction mentioned § 16. When accompanied by sulphide of lead [No. 89], only a small part of the antimony is converted into antimonious acid, which sublimes; the remainder is changed into a white powder consisting of a mixture of antimonate of oxide of antimony, sulphate of lead, and antimonate of lead. When a compound containing sulphide of lead or bismuth, besides sulphide of antimony, is heated on Ch in the R Fl, a Ct is deposited consisting of antimonious acid mixed with sulphate of lead or bismuth, and, nearer to the assay, a yellow one of the oxides of lead or bismuth; how in such a case the presence of antimony may be ascertained v. § 87.

§ 53. To detect a small amount of sulphide of antimony in sulphide of arsenic, Plattner strongly recommends the following method, by which he obtained very decisive and satisfactory results: The assay [No. 88] is introduced into a glass tube, sealed at one end, and gently heated; the sulphide of arsenic is volatilized, and the greater part of the sulphide of antimony remains as a black powder in the lower end of the tube; this end is cut off, the black substance taken out and transferred to a tube open at both ends. By applying heat the characteristic antimony-reaction will appear.

### *Arsenic.*

The reactions of arsenic and its compounds, see § 11, § 15, § 20, § 34, and Table II, 2.

§ 54. All metallic arsenides yield, when heated in the open glass tube, a sublimate of arsenous acid (v. § 15), and most of them evolve a garlic odor (v. § 20) when heated on Ch in R Fl [No. 77].

Some metals, ex. gr. nickel and cobalt, have a great affinity for arsenic, so that, when only a small quantity of the latter is present, the characteristic odor is not observable ; in such cases it is sometimes produced when the metallic compound is fused on Ch with some pure lead in the O Fl.

§ 55. The sulphides of arsenic, heated in the open glass tube, evolve sulphurous acid and yield a sublimate of arsenous acid. To show in a very decisive manner the presence of arsenic in any of its combinations with sulphur, the powdered assay [No. 80] is mixed with six parts of a mixture of equal parts of cyanide of potassium and carbonate of soda, the mass introduced into a tube sealed at one end, and heat applied, at first very gently but gradually raised to redness. A ring of metallic arsenic will be deposited in the colder part of the tube

§ 56. When sulph-arsenides are heated on Ch, the whole of the arsenic, especially when only small quantities are present, may pass off in combination with sulphur; but when such compounds [No. 88] are mixed with from three to four parts of cyanide of potassium and exposed to the R Fl, sulphide of potassium is formed and the arsenic escapes with its peculiar odor.

§ 57. To detect a very small quantity of arsenous acid, the following way may be pursued: a glass tube provided with a small bulb at one end is close above it narrowly drawn out; the assay [No. 38] is introduced into the bulb, and a charcoal splinter placed into the tube; the narrow aperture through which the tube communicates with the bulb prevents the Ch from coming in contact with the substance. The tube is then heated to redness at the place where the charcoal splinter lies, and as soon as this is incandescent, heat is also applied to the bulb. The arsenous acid is volatilized, and its vapors, while passing over the red-hot charcoal, become reduced and deposit a black metallic ring of arsenic in the colder part of the tube. By cutting the tube below the ring and heating this part by the flame of a spirit-lamp, the arsenic is volatilized, thereby emitting its characteristic odor.

§ 58. To show the presence of arsenic in arsenites and arsenates, it will in most cases be sufficient to mix the substance [No. 38] with carbonate of soda and heat it on Ch in R Fl. Sometimes it is necessary to treat the assay with a mixture of carbonate of soda and cyanide of potassium in the manner mentioned, § 55; and in other cases again, where but small quantities of arsenous or arsenic acid are combined with metallic oxides which are readily reduced, recourse must be had to the humid way.

*Bismuth.*

The reactions of bismuth and its compounds, see § 12, § 17, § 22, and Table II, 3.

§ 59. Bismuth when alloyed with other metals, or when as sulphide in combination with other sulphides, is in many cases, and most especially so when accompanied by lead or antimony, not to be detected by the ring which it deposits on Ch. In such a case the assay [No. 49] is treated on Ch until a copious yellow Ct is formed. The Ct is carefully scraped off from the Ch and dissolved in S Ph on platinum wire with the O Fl. The colorless bead is removed from the wire, placed on Ch, a little metallic tin added, and the whole exposed to the R Fl. If bismuth was present, the glass assumes, on cooling, a dark-gray or black color. The oxides of antimony showing the same behavior, the assay, if not quite free from antimony, has to be treated on Ch in the O Fl until the whole of it has been volatilized, and the remaining mass treated on another piece of Ch as above mentioned.

*Boracic Acid.*

§ 60. With many borates, which do not impart to the outer flame the peculiar yellowish-green color [v. § 35], this reaction may be produced by reducing the substance [No. 2] to powder, adding a drop of concentrated sulphuric acid, fastening the mixture into the hook of the platinum wire, and playing on it with the blue cone of the flame.

§ 61. Another way, and by which even a very small quantity of boracic acid in salts and minerals may be detected, is: to reduce the substance to a very fine powder, to mix it with from 3 to 4 parts of a mixture of $4\frac{1}{2}$ parts of bisulphate of potassa and 1 part of fluor-spar, and to knead the whole with a little water into a thick paste. This mass is then fastened to a platinum wire, and exposed to the blue cone of the flame. While the mass enters into fusion fluoboric acid is formed which, on escaping, colors the flame intensely yellowish-green. The reaction appearing sometimes only for a few seconds, the flame should be very attentively watched during the whole time of the experiment.

*Bromine.*

§ 62. Bromides treated with S Ph and oxide of copper on platinum wire, or treated with sulphate of copper on silver foil, show the same reaction as chlorides (v. § 66), with this difference, that the blue color of the outer flame is rather greenish, especially on the edges [No. 16].

§ 63. To discriminate bromides from chlorides more distinctly, the bromide is fused with bisulphate of potassa, both in the anhydrous state, in a small matrass with long neck. Sulphurous acid is evolved, and the matrass is filled with yellow vapors of bromine, characterized by their peculiar odor. The color of the gas is only clearly seen at daylight.

*Cadmium.*

The reactions of cadmium and its compounds, see §§ 11, 24, and Table II, 4.

§ 64. To detect a very small quantity of cadmium, one per cent. or less, in zinc or its ores, the pulverized assay is mixed with Sd and exposed for a short time to the R Fl on Ch. A distinct Ct of oxide of cadmium is deposited. The zinc being less volatile, evaporates only with continued blowing [No. 53].

*Chlorine.*

§ 65. Some oxide of copper is dissolved by means of the O Fl in a bead of S Ph on platinum wire, until it has assumed a deep-green color. Some grains of the pulverized assay [No. 18] are then made to adhere to the bead, and both heated with the blue cone of the flame. If chlorine is present the flame now assumes an intense azure-blue color, owing to the formation of chloride of copper (v. § 36). This test is very delicate, and will show the presence of a very minute quantity of chlorine.

§ 66. Another method is to place on silver-foil some protosulphate of iron, or some sulphate of copper, to moisten it with a drop of water, and then to add the assay [No. 18]. After a while the silver will be found blackened. Substances which are insoluble in water have previously to be fused with a little Sd on platinum wire, to form a soluble chloride [No. 10].

Chlorides, when moistened with sulphuric acid and exposed to the Blp flame, impart to it a faint green coloration which, however, is generally confined to the inner cone, and is quantitatively of much less intensity than that produced with borates. A small amount of boracic acid, when occurring together with a chloride, can, therefore, not be detected by the method mentioned § 60.

### *Chromium.*

§ 67. Oxide of chromium gives very characteristic reactions with the fluxes on platinum wire (v. Table II, 6), but when accompanied by a large quantity of iron, copper, or other substances which also intensely color the Bx and S Ph beads, the chromium color frequently becomes very indistinct.

§ 68. In such a case, and when the chromium is not in combination with silicic acid, its presence may be detected in the following manner: The assay-piece [No. 71] is reduced to a fine powder and mixed with about four times its own volume of a mixture of equal parts of Sd and nitre. The mass is fastened into the hook of a thick platinum wire, or placed into a small platinum spoon, and treated with a powerful O Fl. An alkaline chromate is formed which is dissolved in water, the solution supersaturated with acetic acid, and a crystal of acetate of lead added. If chromium was present, a yellow precipitate of chromate of lead will appear. The precipitate may be collected on a filter and tested in the Bx and S Ph beads, when the characteristic chromium-reactions will be produced.

### *Cobalt.*

The reactions of cobalt, see Table II, 7.

§ 69. To detect cobalt when in combination with other metals, v. § 83.

To show its presence in arsenides, the assay [No. 78] is placed on Ch and heated until no longer fumes of arsenous acid are emitted. (Lead and bismuth, if present, form the characteristic coatings.) Bx is now added and the heat continued until the glass appears colored. If the color is not pure blue, the presence of iron is indicated. The glass is in this case removed from the globule, and the latter treated repeatedly with fresh quantities of Bx until the pure cobalt-color is obtained. Nickel and copper, if present, do not enter into the flux before the whole of the cobalt is oxidized. If we wish to ascertain the presence of these metals, the glass which is colored by cobalt is removed from the globule, and the latter treated with fresh portions of Bx in the O Fl until the color of the bead becomes brown, indicative of nickel. The glass is again removed and the globule treated with S Ph in the O Fl; when copper is present the bead assumes a green color, which remains unaltered on cooling. Treated with tin on Ch the glass turns opaque and red.

§ 70. To detect cobalt in sulphides, the assay [No. 79] is heated on Ch in the R Fl until all volatile substances are driven off, the remaining mass reduced to powder, well calcined, and the calcined mass treated with Bx on Ch in the O Fl. If cobalt is the only coloring metal present, the bead will exhibit a pure blue color; a small addition of iron will make the glass appear green while hot, but blue when cold. Copper and nickel, when present to some extent, will prevent the cobalt-color to be distinctly seen. The bead is in this case exposed to the R Fl until it appears transparent and flows quietly; the oxides of copper and nickel are by this means reduced, and the pure color of cobalt, or that of cobalt mixed with iron, becomes apparent.

## *Copper.*

The reactions of copper and its compounds, see §§ 35, 36, and Table II, 8.

§ 71. The red color which copper imparts to the Bx or S Ph bead, when heated on Ch in the R Fl in contact with tin (v. Table II, 8), is very characteristic and will in most cases clearly show the presence of this metal. But if only a small quantity of copper is associated with other metals, the reaction is not easily obtained; in this case we may proceed as follows:

The assay [No. 89, or No. 86, or No. 85] is placed on Ch and played upon with the O Fl until antimony and other volatile metals are driven off. Some vitrified boracic acid is fused on Ch to a glassy globule, the assay placed close to it, and the whole covered with a large R Fl. When the metallic globule begins to assume a bright metallic surface, the flame is gradually converted into a sharply-pointed blue cone, which is made to act only on the glass, leaving the metallic globule untouched, and so situated that it touches the glass on one side, and on the other side is in close contact with the Ch. During this process lead, iron, cobalt, part of the nickel, and such of the more volatile metals, that were not entirely removed by the previous calcination, as bismuth, antimony, zinc, &c., become oxidized, and their oxides partly volatilized and partly absorbed by the boracic acid. The remaining metallic globule is then removed from the flux and treated on Ch with S Ph in the O Fl, when the copper is oxidized and dissolved. The limpid bead is then re-fused in the R Fl with addition of tin. A trace of copper may thus be made to produce distinctly the characteristic reaction.

§ 72. To show the presence of copper in compounds which contain much nickel, cobalt, iron, and arsenic, the assay [No. 82] is first treated with Bx on Ch in the R Fl, when the greater part of iron and cobalt are dissolved. The remaining globule is then mixed with some pure lead, and treated as shown § 71. Arsenic is for the most part driven off, and the rest of the iron and cobalt, with some nickel, absorbed by the boracic acid. The globule is removed from the glass and treated with S Ph in the O Fl; dark-green while hot, and somewhat lighter green when cold (produced by the mixture of the yellow of nickel and the blue of copper), indicates the presence of copper.

To detect copper when in combination with tin v. § 110.

§ 73. To detect copper in sulphides, the pulverized assay [No. 76] is calcined, and the calcined mass treated as above, or, when the amount of copper is not very small, simply treated with Bx or S Ph on Ch in the O Fl, and subsequently with addition of tin in the R Fl. The presence of copper is then shown by the red color and the opaqueness of the glass on cooling. This reaction is only prevented or, at least, made indistinct by antimony or bismuth, which cause the glass to turn gray or black. In this case the assay is, after calcination, mixed with Sd, Bx, and some pure lead, and the mixture fused on Ch in the R Fl. The metallic globule is then heated on Ch to drive off the antimony, and afterwards treated with boracic acid as above.

§ 74. When a mineral which contains copper is heated in the blue cone, the outer cone of the flame frequently assumes a green or, if the metal is in combination with chlorine, an azure-blue color. This reaction, if not produced by heating the substance alone, may sometimes be elicited by adding a drop of concentrated hydrochloric acid to the pulverized assay [No. 73], evaporating to dryness, mixing the dry powder with a little water to a stiff paste, fastening this into the hook of a platinum wire, and then exposing it to the blue cone of the flame.

### *Fluorine.*

§ 75. To detect fluorine in such minerals where it occurs only as an accessory element in combination with weak bases, and which at the same time contain water, a small piece of the substance [No. 60] is placed into a glass tube sealed at one end, a wet Brazil-wood paper introduced into the open end, and heat applied. Fluoride of silicon and hydrofluoric are evolved; the former is decomposed by the watery vapor and deposits a ring of silica not far distant from the assay, and the latter turns the red color of the test-paper into straw-yellow. Mica, containing not more than $\frac{3}{4}$ per cent. of fluorine shows the reaction very distinctly.

§ 76. To show the presence of fluorine in minerals where it is united with strong bases, the finely powdered assay [No. 6] is mixed with about four parts of bisulphate of potassa and introduced into a glass tube, sealed at one end. Heat is applied until sulphuric acid begins to escape. The sides of the tubes become covered with silicic acid, resulting from the decomposition of the gaseous fluoride of silicon. The tube is cut off close above the fused mass, cleaned with water, and carefully dried with blotting paper. The dulled appearance of the glass indicates the presence of fluorine.

§ 77. Another process, and by which the presence of fluorine in all kind of compounds may be shown, is to mix the pulverized assay with some S Ph which has previously been fused on Ch and then reduced to powder; to place the mixture on platinum foil, which is connected with an open glass tube in such a manner as to constitute a kind of tubular continuation to the former, and to heat with the blow-pipe flame until the mass enters into fusion. If the flame is so directed that the products of decomposition are made to pass through the glass tube and a moistened Brazil-wood paper is introduced into the other end, the presence of hydrofluoric acid is indicated by the change of color which the latter experiences; in some cases the glass will also be dulled, or a deposit of silicic acid be formed. This test is very delicate.

### *Gold.*

§ 78. When gold is in combination with metals which are volatile at a high temperature, ex. gr. tellurium, mercury, antimony, it is only necessary to heat the alloy on Ch with the O Fl, when the gold remains behind in a pure state and may be recognized by its physical properties. Lead is removed by the process of cupellation, as explained in § 102.

§ 79. When associated with copper, the presence of which is easily detected by S Ph on Ch, the alloy, for example gold-coin, is dissolved in pure melted lead and the new compound subjected to the process of cupellation on bone-ash. Copper is by this means entirely removed. To test the remaining globule for silver, it is treated with S Ph on Ch in the O Fl; the silver is gradually oxidized and dissolved by the glass, which when cold assumes an opal-like appearance. To determine approximately the relative proportions of the two metals, the metallic globule is taken from the cupel, placed in a small porcelain dish, containing some nitric acid, and heat applied. If the alloy contains 25 per cent. of gold or less, it turns black, the silver is gradually dissolved and the gold remains behind as a brown or black spungy or pulverulent mass. If the alloy contains more than 25 per cent. of gold, the globule turns also black, but the silver is not dissolved. If both metals are present in about equal proportions, the globule remains unaltered. If the amount of gold is considerable it is indicated by the color of the alloy.

§ 80. When associated with metals, which *per se* are infusible before the blow-pipe, as ex. gr. platinum, iridium, palladium, the metallic globule obtained by cupellation shows much less fusibility than pure gold. The exact nature of the foreign metals cannot be ascertained before the Blp; the humid way must be resorted to.

*Iodine.*

§ 81. Iodides, tested with a S Ph bead which is saturated with oxide of copper as shown § 65, impart to the outer flame a fine green color [No. 17].

Fused with bisulphate of potassa in a glass tube, closed at one end, violet vapors are evolved, iodine sublimes, and sulphurous acid escapes.

§ 82. Another method, which is said to surpass in delicacy even the reaction with starch, is to mix the substance with a mixture of carbonate of lime and quicklime, to dry the mass thoroughly, to add some protochloride of mercury (corrosive sublimate), to rub the whole well together, and to place it in a glass tube closed at one end. The tube is then narrowly drawn out a little above the assay, and the mass heated to redness. Protiodide of mercury is formed, which sublimes in yellow or red needles into the narrow tube. This reaction is founded on the property of lime to decompose the protochloride of mercury, but not the protiodide.

*Iron.*

The reactions of the oxides of iron, see Table II, 10.

§ 83. The colors which iron imparts to the various fluxes are sufficiently characteristic to ascertain its presence in such metallic compounds which contain no easily fusible substances, by simply treating the assay with Bx on Ch in the O Fl. When lead, tin, bismuth, antimony, or zinc are present, the R Fl is employed, and directed in such a manner that it principally touches the glass. Thus, the oxidation and consequent saturation of the bead with the oxides of these metals, is to a great extent prevented. In either case the glass, while still soft, is removed from the globule and exposed on another place of the Ch to the R Fl. Those metals whose oxides are easily reduced, are now precipitated, and the characteristic bottle-green color of iron is clearly observable, unless cobalt be present. In this case the glass is again softened with the R Fl, separated from the precipitated metals, fastened into the hook of a platinum wire and treated with the O Fl until the whole of the iron may be supposed to be converted into sesquioxide. The glass, while hot, will appear green, and blue when cold, if only a trace of iron is present. But when the amount of iron is more considerable, it will be dark-green while hot and bright-green when cold, the latter color resulting from the mixture of the blue of cobalt and the yellow of iron. The metals remaining behind on Ch after the treatment with Bx, and which frequently are only copper and nickel (lead, antimony, and bismuth being volatilized), may be treated as shown § 71.

To detect iron in arsenides and sulphides, the assay is well calcined, and the calcined mass treated as above [No. 86 and No. 79].

§ 84. The oxides of iron when associated with a large quantity of manganese [No. 84 and No. 69], color the Bx bead on platinum wire in the O Fl, red. To show the presence of iron the bead is removed from the wire, placed on Ch, and treated with tin in the R Fl. The vitriol-green color of iron will appear in its purity. When associated with the oxides of manganese and cobalt, a minute quantity of iron cannot very well be detected by means of the blow-pipe alone. When accompanied by the oxides of copper and nickel [No. 78 or No. 85], the assay is dissolved in Bx on Ch in the O Fl and the glass treated as shown § 83.

§ 85. The presence of chromium prevents any conclusive deduction as to the presence of iron from the color of the beads. In such a case the substance [No. 71] may be mixed with three parts of nitre and one of Sd, and the mixture fused in small portions into the hook of a thick platinum wire. The alkaline chromate is dissolved in water and the residue treated with the fluxes. The presence of the oxides of iron when associated with the oxides of uranium cannot be ascertained by means of the blow-pipe alone.

### *Lead.*

The reactions of lead and its compounds, see §§ 12, 23, 36, and Table II, 12.

§ 86. An alloy of lead and zinc [No. 50] deposits a Ct of oxide of lead mixed with oxide of zinc; the presence of lead is shown by the color of the Ct and by the azure-blue tinge which it imparts to the R Fl (v. § 23).

An alloy of lead and bismuth [No 49] deposits a Ct somewhat darker than that of pure lead, in which the presence of bismuth may be detected as shown § 59, and the presence of lead by the azure-blue color of the R Fl.

§ 87. To detect lead in sulphides, the substance is placed on Ch and treated with the R Fl; the lead is detected by its Ct. An admixture of antimony cannot by this means be ascertained, since the ring of sulphate of lead, surrounding that of the oxide, bears a striking resemblance to the Ct formed by antimonious acid. In this case the pulverized assay [No. 85] is mixed with a sufficient quantity of Sd, and treated for a short time with the R Fl. If no antimony is present a pure yellow Ct with bluish-white edges is formed; but in presence of antimony this Ct is surrounded by another, white one, of antimonious acid. The oxide of lead Ct appears, moreover, darker than usual, resembling that of bismuth, owing probably to the formation of antimonate of lead. If this Ct is scraped off from the Ch and treated with S Ph as mentioned § 59, in the case of bismuth, the bead, on cooling, assumes a black color, whereby, in absence of bismuth, the presence of antimony is proved. A very small quantity of antimony can by this method not be found out with certainty, since, by keeping up the blast for some time, the sulphide of sodium begins to vaporize and to coat the Ch with a ring of sulphate of soda (v. § 30).

§ 88. When sulphide of lead is associated with a considerable quantity of sulphide of copper [No. 89], the metallic globule, obtained by the process of reduction, does not betray, by its physical properties, the presence of lead. But if the alloy is removed from the flux and played upon with a powerful O Fl, the greater part of the lead will be volatilized and deposit a Ct.

## *Lithia.*

§ 89. To detect lithia in silicates which contain only little of it, proceed as follows: The substance [No. 67] is reduced to a fine powder and mixed with about 2 parts of a mixture of 1 part of fluor-spar with $1\frac{1}{2}$ parts of bisulphate of potassa; a few drops of water are added and the whole kneaded into a paste. The mass is fused with the blue cone of the flame into the hook of a platinum wire. If lithia is present the outer flame will appear red. The color is not very intense, and verging into violet. The presence of potassa does not prevent the reaction, but makes the flame appear still more violet; soda makes the reaction uncertain.

*Manganese.*

The reactions of manganese, see Table II, 13.

§ 90. The presence of manganese in any compound substance is readily detected by mixing the pulverized assay [No. 66 or No. 84] with about 2 or 3 parts of Sd, and fusing it by means of the O Fl on platinum foil. Manganate of soda is formed, which, while hot, is green and transparent, and, on cooling, turns bluish-green and opaque. The reaction is very distinct when as much as one-tenth per cent. of manganese is present. But even the slightest trace may be detected when, instead of Sd, a mixture of 1 part of nitre with 2 parts of Sd is used. Chromium does not prevent the reaction, merely changing the color to yellowish-green. It is only in presence of silica and cobalt that this test is not available, since at a high temperature the silica unites with the soda to silicate of soda, which, in dissolving the oxide of cobalt, produces a blue glass, and thus interferes with the manganese color.

*Mercury.*

The reactions of mercury and its compounds, see §§ 11, 17, and Table II, 14.

§ 91. Mercury is detected in amalgams [No. 47] by the sublimate of metallic mercury which they yield, when heated in a glass tube closed at one end.

When in combination with sulphur [No. 81], chlorine [No. 39], iodine or ox-acids, the substance is previously mixed with some anhydrous Sd or some neutral oxalate of potassa. The acids, &c., are retained by the soda, and mercury sublimes.

If the quantity of mercury is so small, that the nature of the sublimate cannot with certainty be ascertained, the experiment has to be repeated, a piece of iron wire around which a gold-leaf has been wrapped being at the same time introduced into the tube and held close above the assay. The gold-leaf will turn white if ever so little mercury be present.

*Nickel.*

The reactions of nickel, see Table II, 16.

§ 92. To detect nickel in metallic compounds which are fusible before the Blp, the assay is treated with Bx on Ch in the R Fl; iron, cobalt, &c., enter into the flux and may be detected as shown § 69, while the metals whose oxides are easily reduced remain behind. This operation is repeated until the glass appears no longer colored. The remaining globule is treated with S Ph in the O Fl. We now obtain either the pure color of nickel, or that of nickel mixed with copper (v. § 72); in this case it is treated on Ch with tin, whereby the presence of copper may be ascertained. Bismuth or antimony prevents the reaction for copper, the bead turning black, instead of red. Such compounds must, previous to their treatment with fluxes, be heated on Ch in R Fl until all volatile substances are driven off [No. 82].

In arsenides and sulphides nickel is detected by the methods given for cobalt under the same circumstances (v. § 70).

*Nitric acid.*

§ 93. The perfectly dry substance [No. 23] is heated in a matrass with some bisulphate of potassa; orange-yellow vapors of nitrous acid are emitted, even if but a small quantity of a nitrate is present.

*Phosphoric acid.*

§ 94. A very minute quantity of phosphoric acid may be detected by pulverizing the substance [No. 14], adding a drop of concentrated sulphuric acid, fastening the paste into the hook of a platinum wire, and playing upon it with the blue cone of the flame; the outer flame will assume a bluish-green color (v. § 35).

Certain azotized compounds, as nitric acid, nitrate of ammonia, chloride of ammonium, &c., when fastened into the hook of a platinum wire and touched with the cone of the blue flame, impart to the outer flame a bluish-green color, resembling that caused by phosphoric acid.

§ 95. In a substance, containing not less than about 5 per cent. of phosphoric acid, the presence of the latter may be shown by dissolving the assay [No. 68] on Ch in boracic acid and forcing into the glass, when a good fusion is effected, a piece of fine steel wire; a good R Fl is then given. The iron is oxidized at the expense of the phosphoric acid, causing the formation of a borate of the oxide of iron and phosphide of iron, which fuses at a sufficiently high temperature. The bead is then taken from the Ch, enveloped in a piece of paper, and struck lightly with a hammer, by which means the phosphide of iron is separated from the surrounding flux. It exists as a metallic-looking button, attractable by the magnet, frangible on the anvil, the fracture having the color of iron. If the substance under assay contained no phosphoric acid, the iron wire will keep its form and metallic lustre, excepting at the ends, where it will be oxidated and burnt. The substance to be assayed ought not to contain sulphuric acid, arsenic acid, or any metallic oxides reducible by iron.

Phosphate of lead exhibits the peculiarity of crystallizing on cooling after having been fused on Ch; the crystals have frequently large facets of a pearly lustre.

*Potassa.*

§ 97. The violet color of the flame is sufficiently characteristic for potassa (v. § 33). But being altogether prevented or, at least, made very indistinct by the addition of a few per cent. of soda or lithia, it can only in a very few cases be made use of. For the detection of potassa in silicates it is almost entirely unavailable, because these compounds almost always contain some soda.

§ 98. If the base of a compound consists essentially of potassa, the following method may be advantageously employed for its detection: Some Bx, to which a little boracic acid has been added, is melted into the hook of a platinum wire and so much protoxide of nickel added that the glass on cooling shows a distinct brownish color. A small piece of the substance under examination [No. 15] is made to adhere to the glass and the whole fused together with the O Fl. If the assay-piece contained no potassa, the color of the glass, after perfect cooling, will have remained unchanged; but if potassa was present in sufficient quantity, the glass will appear bluish.

## *Selenium.*

§ 99. The reactions of selenium are very characteristic. In non-volatile compounds, which do not give the red sublimate mentioned § 11, the selenium is detected by heating a small piece of the substance [No. 87] on Ch in O Fl, when the peculiar odor is evolved; if much selenium is present, a Ct is deposited, v. § 28. Selenites and selenates are treated on Ch with Sd in R Fl, when a reduction takes place and the selenium vaporizes with the characteristic odor.

## *Silica.*

§ 100. Pure silica [No. 54], when treated with Bx on platinum wire, dissolves slowly to a transparent glass which fuses with difficulty. Treated with S Ph in the same manner, only a small quantity is dissolved, the rest floating in the liquid bead as a semi-transparent mass. The behavior to Sd see § 39. With a little So Co it assumes a pale bluish color which, on addition of a large quantity of the reagent, turns dark-gray or black; very thin splinters may be fused by a great heat to a reddish-blue glass.

§ 101. Silicates [No. 61], when treated with S Ph on platinum wire, are decomposed; the bases unite with the free phosphoric acid to a transparent glass in which the silica may be seen floating as a gelatinous cloudy mass. The bead ought to be carefully observed while hot, since many silicates form a glass which on cooling opalizes or becomes opaque, when, of course, the phenomenon can no longer be seen. The experiment is best performed with a small splinter of the substance under examination, and only when this does not appear to be affected by the flux, the finely pulverized substance should be used. If but a very small quantity of silica is present, the glass will appear perfectly transparent. Its presence in this case cannot be detected by means of the Blp.

§ 102. Silicates containing at least so much silica that the quantity of oxygen in the acid is twice that of the oxygen in the base, dissolve, when treated with Sd on Ch, with effervescence to a transparent glass which remains so when cold. When less silica is present decomposition also takes place, but the glass turns opaque on cooling, the amount of silicate of soda which is formed not being sufficient to dissolve the eliminated bases.

*Silver.*

The reactions of silver, see § 27, and Table II, 20.

§ 103. When in combination with metals which are volatile at a high temperature, ex. gr. bismuth, lead, zinc, antimony, the substance is heated alone on Ch, when, after evaporation of the foreign metals, a button of pure silver remains behind and a feeble reddish Ct is deposited on the Ch. If associated with much lead or bismuth, these metals are best removed by cupellation, a process which is executed in the following manner: Finely pulverized bone-ash is mixed with a minute quantity of soda and made with a little water into a stiff paste; a hole is now bored into the Ch, filled with the paste, and its surface smoothed and made slightly concave by pressing on it with the pestle of the little agate mortar. The mass is then dried by the flame of a common spirit lamp. On this little cupel the assay [No. 51] is placed and so long heated with the O Fl until the whole of the lead or bismuth is oxidized and absorbed by the cupel. The silver or, if gold is present, the alloy of silver and gold remains as a bright metallic button on the cupel.

§ 104. When combined with metals which are not volatile, but which are easier oxidized than silver, the presence of this metal may in some cases be detected by simply treating the alloy with Bx or S Ph on Ch. Copper, nickel, cobalt, &c., become oxidized and their oxides dissolved by the flux, while silver remains behind with a bright metallic surface. But when these metals are present to a considerable extent, another course has to be pursued, a course which may always be taken when a substance is to be assayed for silver, or silver and gold.

§ 105. The assay-piece [No. 86] is reduced to a fine powder, mixed with vitrified Bx and metallic lead (the quantities of which altogether depend upon the nature of the substance, and for which, therefore, no general rule can be given), and the mass placed in a cylindrical hole of the Ch. A powerful R Fl is given until the metals have united to a button, and the slag appears free from metallic globules. The flame is now converted into a O Fl and directed principally upon the button. Sulphur, arsenic, antimony, and other very volatile substances, are volatilized; iron, tin, cobalt, and a little copper and nickel become oxidized and are absorbed by the flux; silver and gold and the greater part of copper and nickel remain with the lead (and bismuth, if present). When all volatile substances are driven off, the lead begins to become oxidized, and the button assumes a rotary motion; at this period the blast is discontinued, the assay is allowed to cool, and when perfectly cold the lead button is separated from the glass by some slight strokes with a hammer. It is now placed on a cupel of bone-ash and treated with the O Fl until it again assumes a rotatory motion. If much copper or nickel is present, the globule becomes covered with a thick infusible crust, which prevents the aimed-at oxidation; in this case another small piece of pure lead has to be added. The blast is kept up until the whole of the lead and other foreign metals, viz., copper and nickel, are oxidized; this is indicated by the cessation of the rotatory movement, if only little silver is present, or by the appearance of all the tints of the rainbow over the whole surface of the button, if the ore was very rich in silver; after a few moments it takes the look of pure silver. The oxides of lead, copper, &c., are absorbed by the bone-ash, and pure silver, or an alloy of silver with other noble metals, remains behind; the button may be tested for gold, &c., after the method given in § 79.

*Sulphur.*

§ 106. The presence of sulphur in sulphides may in many cases be detected by heating in a glass tube (v. §§ 11, 14), or on Ch with the O Fl.

§ 107. A very delicate test for the presence of sulphur, in whatever combination it may be contained in the substance, and which possesses moreover the advantage over all other methods of being very easily performed, is to mix the pulverized assay [No. 4] with some pure Sd or, better still, with a mixture of two parts of Sd and 1 of Bx, and to treat it on Ch with the R Fl. The fused mass is removed from the Ch, powdered, the powder placed on a silver foil or a bright silver coin, and a drop of water added. If the substance under examination contained any sulphur, a black spot will be formed on the silver foil, owing to the formation of sulphide of silver from the decomposition of the sulphide of sodium, which, in its turn, resulted from the decomposition of the sulphide or sulphate, or other sulphur-compound of the assay-piece, under the influence of Sd, Ch, and a high temperature. Selenium shows the same reaction; it is readily recognized by the peculiar odor which it emits when heated on Ch alone.

§ 108. To decide whether the reaction obtained in the experiment was owing to the presence of a sulphide or to that of a sulphate, the finely-pulverized substance [No. 76] is fused in a small platinum spoon with some hydrate of potassa. The spoon with the contents is then placed into a vessel containing some water, and a piece of silver foil inserted into the liquid. If the silver remains perfectly bright, a sulphate was present, if it turns black, a sulphide. The absence of substances which might exercise a reducing influence is required.

### *Tellurium.*

§ 109. The presence of tellurium in mineral substances is detected by the tests given §§ 11, 18, 29. In presence of lead or bismuth the reactions in the open tubes and on Ch are not quite pure. In this case we may subject the assay to the following treatment: The substance is mixed with some Sd and charcoal-powder, the mixture introduced into a glass tube closed at one end, and heated to fusion; after cooling, a few drops of hot water are poured into the tube; if tellurium was present, telluride of sodium has been formed, which dissolves in hot water with a purplish-red color. This test is applicable to show the presence of tellurium in a great many compounds, even in such where it occurs in the oxidized state.

### *Tin.*

The reactions of tin and its compounds, see §§ 12, 26, 45, and Table II, 22.

§ 110. The presence of tin is indicated by its Ct when the substance [No. 13], alone or mixed with Sd, is exposed to the R Fl on Ch.

When the substance under examination is an alloy, a little Bx is conveniently added, which absorbs the oxide of tin in the measure as it is formed, and allows the presence of those metals which are more volatile, ex. gr. antimony, lead, bismuth, to be recognized by their coatings. Arsenic is detected by its odor, and iron by the color which the Bx bead assumes when re-fused on platinum wire in the O Fl.

To detect copper in tin or its alloy, the assay [No. 52] is fused with a flux consisting of 100 parts of Sd, 50 of vitrified Bx, and 30 of silica. The flame is so directed that the metallic globule assumes a rotatory motion. When in this state the glass is kept covered, as much as possible, with the O Fl, care being taken that the globule is at one side in contact with the glass, and at the other with the Ch. The tin becomes oxidized and the oxide, in the measure as it is formed, absorbed by the flux; the remaining button is copper, pure or with a small quantity of tin, and may be readily tested with the usual fluxes.

*Titanium.*

§ 111. Titanic acid, when forming the principal constituent of any mineral substance, is easily detected by its behavior with the fluxes, v. Table II, 23; but when in combination with bases these reactions are not always clearly perceptible, being frequently suppressed by the predominating reaction of the base. In such case we may subject the assay to the following treatment, by which even very small quantities of titanic acid will become apparent: the substance [No. 65] is reduced to a very fine powder, mixed with from 6 to 8 parts of bisulphate of potassa, and fused in a platinum spoon at a low red-heat; the fused mass is dissolved in a porcelain vessel in the smallest possible quantity of water, aided by heat. There remains an insoluble residue which is allowed to settle; the clear liquid is poured off into a larger vessel, mixed with a few drops of nitric acid and at least six volumes of water, and heated to ebullition. If the substance under examination contained any titanium, a white precipitate of titanic acid forms on boiling. The precipitate is collected on a filter, washed with water, acidulated with nitric acid, and tested with S Ph.

*Uranium.*

§ 112. The presence of this metal is easily recognized, in substances which contain no other coloring constituents, by the reactions given Table II, 25; the most characteristic test is that with

S Ph. In presence of much iron this reaction becomes indistinct; we may then operate in the following manner: the finely-pulverized substance [No. 70] is fused with bisulphate of potassa, the fused mass dissolved in water, mixed with carbonate of ammonia in excess, the liquid separated from the precipitate by filtration, and the filtrate heated to ebullition. If any uranium was present, a yellow precipitate is thrown down, which gives with the fluxes the pure reactions of uranium.

### *Zinc.*

The reactions for zinc and its compounds, see §§ 12, 25, 45, and Table II, 27.

§ 113. A small amount of zinc, when associated with considerable quantities of lead, or bismuth, or antimony, or tin, cannot with certainty be ascertained by means of the Blp.

If the substance under examination contains the zinc as oxide [No. 36], or but a small quantity of sulphide, it is mixed with Sd and treated on Ch in R Fl. Substances consisting essentially of sulphide of zinc may be thus treated without the addition of Sd, and such as contain, beside oxide of zinc, other metallic oxides, are conveniently mixed with some Sd to which about one-half of its weight of Bx has been added. A ring of oxide of zinc is deposited on the Ch. When lead is present [No. 51] the Ct is frequently not pure, being mixed up with the Ct of lead. In this case it is moistened with some So Co and heated again with the O Fl. The oxide of lead is reduced by the red-hot Ch and volatilized, while the oxide of zinc remains behind with a green color (v. § 45).

# FOURTH CHAPTER.

## CHARACTERISTICS OF THE MOST IMPORTANT ORES; THEIR BEHAVIOR BEFORE THE BLOW-PIPE, AND TO SOLVENTS.

§ 114. Of the physical properties of the minerals which are treated of in this chapter, only those are enumerated which serve best to discriminate the different ores from each other. For a more detailed description I must refer to Dana's and other works on mineralogy. Among the distinguishing characters of minerals, their hardness and specific gravity stand foremost. The latter cannot be ascertained without a good balance, and will, for this reason, be of much less use to the practical man than the determination of hardness, an operation which may be performed in a few moments. A set of minerals, representing the scale of hardness, being not always at hand, it will be useful to give a series of substitutes for them, as arranged by Mr. Chapman:

1. Yields easily to the nail.
2. Yields with difficulty to the nail, or merely receives an impression from it. Does not scratch a copper coin.
3. Scratches a copper coin; but is also scratched by it, being of about the same degree of hardness.
4. Not scratched by a copper coin; does not scratch glass.
5. Scratches glass, though rather with difficulty, leaving its powder on it. Yields readily to the knife.
6. Scratches glass easily. Yields with difficulty to the knife.
7. Does not yield to the knife. Yields to the edge of a file, though with difficulty.

8. 9. 10. Harder than flint.

The scale of hardness, as introduced by Mohs, and enlarged by Breithaupt, is as follows:

1. Talc; common laminated light-green variety.
2. Gypsum; a crystalline variety.

2.5. Foliated Mica.
3. Calcareous Spar; transparent variety.
4. Fluor Spar; crystalline variety.
5. Apatite; transparent variety.
5.5. Scapolite; crystalline variety.
6. Orthoclase; white cleavable variety.
7. Quartz; transparent.
8. Topaz; transparent.
9. Sapphire; cleavable varieties.
10. Diamond.

To test the hardness of a mineral we may proceed in two different manners: firstly, by attempting to scratch it with the minerals enumerated in the scale, successively, or, secondly, by abrasion with a file. If the file abrades the mineral under trial with the same ease as No. 4, and produces an equal depth of abrasion with the same force, its hardness is said to be 4. If with more facility than 4, but less than 5, the hardness may be $4\frac{1}{4}$ or $4\frac{1}{2}$. Several successive trials should be made to obtain certain results; and, when practicable, both methods should be employed.

## Ores of Antimony.

### *Gray Antimony* [Stibnite].

§ 115. Sb $S^3$. H=2. G=4.5. Of lead-gray color and metallic lustre. Usually of columnar structure, consisting of a vast number of needle-shaped crystals, sometimes side by side, sometimes divergent. Very brittle.

It fuses readily in the flame of a candle. In a matrass, sometimes yields a slight sublimate of sulphur; on increasing the heat by application of the Blp flame, a sublimate is produced which after cooling is brownish-red, and which consists of a mixture of tersulphide of antimony with antimonious acid. In an open glass tube, emits sulphurous acid and antimonial fumes. On Ch it is volatilized, covering the Ch with oxide of antimony, which, when touched with the R Fl, disappears with a pale greenish-blue tinge.

When pure, wholly soluble in heated hydrochloric acid with evolution of sulphuretted hydrogen; usually a residue of chloride of lead is left. Partly decomposed by caustic potassa; the solution, when mixed with an acid, affords a yellowish-red precipitate.

*Berthierite.*

§ 116. Composition variable, sometimes $FeS+SbS^3$. H=2—3. G=4—4.3. Metallic lustre, less splendent than gray antimony; color dark steel-gray.

Heated in a matrass, fuses and yields a slight sublimate of sulphur; on application of a strong heat, a black sublimate of sulphide of antimony is formed, which, on cooling, becomes brownish-red. In an open glass tube it behaves like the preceding ore. In Ch, fuses easily and coats the charcoal with oxide of antimony; there remains, finally, a black slag, which is attracted by the magnet and gives with fluxes the iron reaction.

Soluble in hydrochloric acid.

*Red Antimony* [Kermesite].

§ 117. $2\,SbS^3+SbO^3$. H=1—1.5. G=4.5—4.6. Usually in tufts of capillary crystals of cherry-red color.

In a matrass, fuses readily and yields a slight yellowish-red sublimate; with strong heat, boils and gives a black sublimate which, when cold, is brownish-red. In an open tube and on Ch, behaves like gray antimony.

It dissolves in hydrochloric acid with evolution of sulphuretted hydrogen. The powdered mineral, when treated with caustic potassa, assumes an ochre-yellow color and dissolves completely.

## Ores of Arsenic.

*Native Arsenic.*

§ 118. As, with traces of Sb, Ag, Fe, Co, and Ni. H=3.5. G=5.9. Of metallic lustre and tin-white color, tarnishing on exposure to air to dark-gray.

Heated in a matrass, sublimes; on Ch, behaves like pure arsenic. In both cases, sometimes, a residue is left, which, when treated with fluxes, exhibits the reactions of iron, cobalt, and nickel. (See § 83.)

*Realgar.*

§ 119. $As\,S^2$. H=1.5—2. G=3.4—3.6. Usually of bright-red, sometimes of orange-yellow color, and resinous lustre. Sectile.

In a matrass, fuses, boils, and finally sublimes; the sublimate, after cooling, is red and transparent. In an open glass tube, when carefully heated, yields a sublimate of arsenous acid, sulphurous acid escaping. On Ch, fuses readily and burns with a yellowish-white flame, emitting grayish-white fumes which possess the peculiar alliaceous odor. Subjected to the treatment described § 55, a sublimate of metallic arsenic is obtained.

Not easily affected by acids; but aqua regia dissolves it with continued digestion, part of the sulphur being precipitated. A heated solution of caustic potassa decomposes it, leaving a brownish-black powder ($As^6S$) undissolved.

*Orpiment.*

§ 120. $AsS^3$. H=1.5—2. G=3.4. A foliaceous mineral of lemon-yellow color, and resinous or pearly lustre. Sectile.

Before the Blp, behaves like the preceding, with this difference, that the sublimate, after cooling, is dark yellow and transparent.

Soluble in aqua regia, caustic potassa, and ammonia.

*White Arsenic* [Arsenolite].

§ 121. $AsO^3$. H=1.5. G=3.6. Occurs usually in minute capillary crystals of a white color, and vitreous or silky lustre.

Before the Blp it behaves like pure arsenous acid (v. §§ 9, 15, Table II, 2).

Slightly soluble in hot water; more so in water acidulated with hydrochloric acid.

## Ores of Bismuth.

*Native Bismuth.*

§ 122. Bi; H=2—2.5. G=9.7. Color silver-white, tinged with red. Lustre metallic. Brittle when cold; but, when hot, may be laminated. Occurs foliated, granular, and arborescent; occasionally crystallized.

Before the Blp it behaves like pure bismuth (v. §§ 17, 22).

Readily dissolved by nitric acid; the solution is precipitated by water.

*Telluric Bismuth* [Tetradymite].

§ 123. Bi and Te in variable proportions. H=1.5—2. G=7.2

—8.4. Of pale steel-gray color, and high metallic lustre. Occurs usually in tabular crystals, or foliated masses; the laminæ are elastic. It soils paper.

In an open glass tube it fuses readily, emitting a white smoke which partly condenses, coating the tube near the assay-piece with a white powder, intermixed with red spots; on directing the flame on this Ct, it fuses to colorless drops ($TeO^2$), while the red sublimate (Se) disappears. On Ch, fuses instantly to a metallic globule which, when touched with the inner flame, imparts a bluish-green color to the outer one, sometimes gives out selenium vapors, and deposits, close to the assay-piece, a dark orange Ct, surrounded at a greater distance by a white Ct.

Soluble in nitric acid.

### *Bismutite.*

§ 124. $3(BiO^3.CO^2+HO)+BiO^3.HO$. H=4—4.5. G=6.9. Usually of a white or light greenish color, and vitreous lustre; in acicular crystallizations.

In a matrass, decrepitates, yields a little water, and turns gray. On Ch, fuses very readily and is reduced, with effervescence, to a metallic globule, covering the Ch with a Ct of oxide of bismuth. If the blast is kept up for some time the whole of the bismuth is volatilized and there remains a scoriaceous mass which, in the R Fl, may be fused to a globule, and which with fluxes gives the indications of copper and iron. With Sd it usually gives the sulphur reaction (§ 107).

Dissolves in hydrochloric acid with effervescence; the solution has a yellow color.

### *Bismuthine.*

§ 125. $BiS^3$. H=2—2.5. G=6.4—6.55. In acicular crystals or massive; of metallic lustre, and lead-gray color, with a yellowish or iridescent tarnish.

In a matrass, fuses and yields a slight sublimate of sulphur. Carefully heated in an open tube, it fuses and yields sulphurous acid and a coat of sulphate of bismuth; the latter may be fused, by application of the Blp flame, to brown drops which, when cold, appear yellow and opaque. On Ch, fuses and boils, throwing out small drops in a state of incandescence, and deposits a Ct of oxide of bismuth.

Soluble in nitric acid with deposition of sulphur. The solution gives a white precipitate with water.

*Bismuth Ochre.*

§ 126. $BiO^3$, containing minute quantities of $Fe^2O^3$, CuO, and $AsO^5$. G=4.36. Occurs usually pulverulent or earthy.

Before the Blp it behaves like pure oxide of bismuth. Soluble in nitric acid.

### Ores of Chromium.

*Chromic Iron.*

§ 127. (FeO, CrO, MgO)+($Cr^2O^3$, $Al^2O^3$). H=5.5. G=4.3—46. Occurs usually massive; of iron-black or brownish-black color, with a shining and somewhat metallic lustre. Some varieties are magnetic.

Heated in a matrass, remains unchanged. Infusible in the forceps. After having been exposed to the R Fl it follows the magnet. In Bx and S Ph slowly, but completely, soluble to a transparent glass, which is beautiful green after cooling. Mixed with Sd and nitre and heated on platinum-foil, the mass fuses and becomes yellow. With Sd on Ch in R Fl it affords metallic iron.

Concentrated acids affect it but little, even when finely pulverized; they dissolve only a little iron. Fused with caustic potassa, chromate of potassa is formed.

### Ores of Cobalt.

*Smaltine.*

§ 128. (Co, Fe, Ni) As.H=3.5—6. G=6.4—7.2. Of tin-white or steel-gray color, and metallic lustre.

In a matrass, usually yields, when heated to redness, a sublimate of metallic arsenic. In an open glass tube, affords a copious sublimate of crystallized arsenous acid, and sometimes emits sulphurous acid. On Ch it fuses readily, with emission of copious arsenical fumes, to a grayish-black magnetic globule which, with the fluxes, gives the indications of iron, cobalt, and nickel.

With nitric acid it gives a pink solution, arsenous acid being deposited.

### *Cobaltine.*

§ 129. $CoS^2+CoAs$. H=5.5. G=6—6.3. Of silver-white and sometimes reddish color, and metallic lustre.

Unchanged in the matrass. In an open glass tube, yields a sublimate of arsenous acid and vapors of sulphurous acid. On Ch, emits copious arsenical and sulphur fumes and fuses to a dull black metallic globule, which is attracted by the magnet, and which, when treated with fluxes, gives the indications of cobalt and iron, and sometimes also those of nickel.

Dissolves in heated nitric acid, arsenous acid being deposited.

### *Cobalt Pyrites* [Linnæite].

§ 130. $CoS+Co^2S^3$. H=5.5. G=4.8—5. Of a more or less bright steel-gray color, and metallic lustre. Crystallizes in the regular octahedron.

In an open glass tube, sulphurous acid is abundantly evolved and sometimes a slight sublimate of arsenous acid formed. On Ch, small pieces of the mineral readily fuse to a globule which, when cold, is covered with a black rough crust, and which is attracted by the magnet. The pulverized mineral, after having been well calcined, dissolves in Bx in O Fl to a blue transparent glass. In a highly saturated bead of this kind, when treated on Ch with R Fl, particles of metallic nickel may be seen floating about.

Soluble in nitric acid, excepting the sulphur.

### *Cobalt Bloom* [Erythrine].

§ 131. $3CoO. AsO^5+8HO$. H=1.5—2.5. G—2.9. Usually of crimson or peach-red color; when crystallized, of pearly lustre; frequently dull and earthy, forming incrustations.

Heated in a matrass, loses water, and the color changes to blue or green. A small crystal, exposed to the inner flame, fuses and colors the outer flame pale-blue. On Ch in R Fl, emits arsenical fumes and melts to a dark-gray globule of arsenide of cobalt which, with fluxes, gives the pure cobalt-reactions.

Acids dissolve it readily to a rose-colored liquid; the solution in concentrated hydrochloric acid appears blue, while hot. The pulverized mineral is partly decomposed by caustic potassa; the powder assumes a bluish-gray color and the solution is sapphire-blue.

*Lavendulan.*

§ 132. $AsO^5$, CoO, NiO, CuO, and HO. H=2.5—3. G=3. Amorphous, with a greasy lustre; color lavender-blue.

Heated in a matrass, gives out water. In the forceps, fuses easily and colors the outer flame pale-blue; the fused mass becomes crystalline on cooling. On Ch in R Fl it fuses with emission of arsenical fumes. With fluxes, gives the reactions of Co, Ni, and Cu (see § 92).

*Earthy Cobalt.*

§ 133. It is a variety of Wad (see § 184), containing sometimes a considerable quantity of oxide of cobalt, in combination with silicic or arsenic acid.

With Bx in O Fl, gives a dark-violet glass, which in the R Fl becomes blue. The S Ph bead when treated on Ch with metallic tin frequently exhibits the copper-reaction. With Sd on platinum-foil it shows the presence of manganese.

Soluble in hydrochloric acid with evolution of chlorine; the solution is usually blue, and on addition of water becomes red.

## Ores of Copper.

*Native Copper.*

§ 134. Pure Copper. H=2.5—3. G=8.9. Of metallic lustre, and copper-red color. Occurs usually massive or arborescent.

It fuses on Ch to a globule which, if the heat is sufficiently high, assumes a bright bluish-green surface; on cooling it becomes covered with a crust of black oxide. With the fluxes it gives the usual indications of copper.

It dissolves readily in nitric acid.

*Copper Pyrites* [Chalcopyrite].

§ 135. $Cu^2S+Fe^2S^3$. H=3.5—4. G=4.1—4.3. Of a brass-yellow color and metallic lustre; on exposure to moist air it becomes iridescent on its surface. It occurs crystallized, but usually massive. It is easily scratched with a knife, giving a greenish-black powder.

Heated in a matrass, decrepitates and yields sometimes a faint sublimate of sulphur, assuming at the same time a darker color or becoming iridescent. Heated in an open glass tube, sulphurous acid is given out abundantly. On Ch, when heated, it blackens, but becomes red on cooling; with continued heat it fuses to a black globule, which is attracted by the magnet; this globule is brittle and reddish-gray in the fracture. The pulverized mineral, after roasting, gives with fluxes the indications of iron and copper. With Sd on Ch it is reduced; the metals are obtained in separate masses. Moistened with hydrochloric acid it colors the flame blue, even previous to fusion.

It dissolves in nitric acid and, more readily, in aqua regia, leaving a residue of sulphur.

### *Purple Copper* [Erubescite].

§ 136. 3 $Cu^2S+Fe^2S^3$. H=3. G=4.4—5. When crystalline, it usually affects the cubical form, and is of a pale yellowish color; when massive, its color is copper-red to reddish-brown; it speedily tarnishes, assuming various hues, mostly purple, blue, and reddish. When scratched with a knife it gives a grayish powder.

Before the Blp it shows pretty much the same behavior as copper pyrites.

Concentrated hydrochloric acid dissolves it, leaving the greater part of the sulphur behind.

### *Copper Glance.*

§ 137. $Cu^2S$. H=2.5—3. G=5.5—5.8. Of a blackish lead-gray color, often with a bluish or greenish tint on its surface. Occurs usually in compact masses, very often shining.

Heated in a matrass, nothing volatile is given out. In an open tube, sulphurous acid is evolved. On Ch, readily fuses to a globule, which boils, and emits glowing drops, sulphurous acid escaping abundantly; the outer flame is at the same time colored blue. With Sd on Ch it yields a globule of metallic copper.

In heated nitric acid it dissolves, leaving a residue of sulphur.

### *Gray Copper* [Tetrahedrite].

§ 138. 4 ($Cu^2S$. FeS. ZnS) ($SbS^3$. $AsS^3$) frequently containing silver and mercury. H=3—4.5. G=4.5—5. Color between steel-gray and iron-black.

Heated in a matrass, fuses and finally yields a dark-red sublimate of tersulphide of antimony with antimonious acid. In an open glass tube, fuses and gives thick fumes of antimony (and arsenous acid), and sulphurous acid; mercury, when present, condenses in the upper part of the tube, forming a metallic mirror. On Ch it fuses readily to a globule, emitting thick white fumes and sulphur vapor; coatings of antimonious acid and of oxide of zinc are deposited; the latter is nearer to the assay-piece and may be tested with SoCo [v. § 45]. To detect arsenic, v. § 56. To detect mercury, add to the finely pulverized assay three times its weight of dry Sd and treat the mixture as directed § 91. The pulverized mineral, after having been well roasted, gives with the fluxes the indications of iron and copper; with Sd, affords metallic copper and a little iron. To detect silver, treat the mineral with pure lead and Bx as directed § 105.

When pulverized it is decomposed by nitric acid, the solution has a brownish-green color; antimonious acid (and arsenous acid) and sulphur remain undissolved. Caustic potassa effects partial decomposition; the sulphide of antimony (and arsenic) enters into solution, and is, on addition of an acid, re-precipitated.

*Tennantite.*

§ 139. $4(Cu^2S, FeS), AsS^3$. H=3.5—4. G=4.37—4.5. Always crystallized; metallic lustre; color blackish lead-gray to iron-black.

In a matrass, gives a sublimate of tersulphide of arsenic. In an open tube, sulphurous acid and a sublimate of arsenous acid. On Ch, fuses easily with emission of sulphur and arsenic vapors to a dark-gray globule, which is attracted by the magnet. The pulverized mineral gives, after calcination, with fluxes, the reaction of iron and copper.

*Arsenical Copper* [Domeykite].

§ 140. $Cu^6As$. H=3—3.5. Reniform, massive, or disseminated; lustre metallic; color tin-white; black and soft when impure.

Heated in a matrass, yields a little water and a sublimate of arsenous acid; the assay-piece assumes a silver-white color. In an open tube, affords a crystalline sublimate of arsenous acid. On Ch, fuses easily with emission of a strong alliaceous odor to a yellowish metallic mass, which gives the copper reactions.

Readily soluble in nitric acid; decomposed by hydrochloric acid, metallic arsenic remaining undissolved.

*Atacamite.*

§ 141. $CuCl+3CuO+3HO$. H=3—3.5. G=4—4.3. Occurs crystalline, or massive lamellar; color various shades of bright green, sometimes blackish-green.

Heated in a matrass, gives out water and a gray sublimate, which, on cooling, becomes grayish-white; the water shows acid reaction. On Ch, fuses readily, colors the outer flame azure-blue, and is finally reduced to a globule of metallic copper; two coatings are deposited on the Ch, the one grayish-white, and the other brownish, which, on being played upon with the R Fl, change their place with an azure-blue tinge.

Easily soluble in acids.

*Red Copper.*

§ 142. $Cu^2O$. H=3.5—4. G=5.8—6. Usually of a very intense, deep red color, occasionally crimson-red; exceedingly friable.

Heated in the pincers, fuses and colors the outer flame emerald-green; moistened with hydrochloric acid and treated in the same manner, the color is azure-blue. On Ch it blackens, then fuses quietly, and finally yields a globule of metallic copper which, on cooling, becomes covered with a coating of black oxide.

Dissolves readily in nitric acid. With hydrochloric acid it gives a brownish solution, which on addition of water is decomposed, a white precipitate of subchloride of copper being formed. It is also soluble in ammonia: the solution is colorless when the access of air is prevented; on exposure to air it turns blue.

*Malachite.*

§ 143. $2CuO.CO^2+HO$. H=3.5—5. G=3.7—4. Occurs usually in the shape of mammillated concretions; the interior is very compact, and lustre shining, in the fracture sometimes earthy, sometimes silky; of a bright green color.

Heated in a matrass, gives out water and turns black. On Ch fuses to a globule, and affords metallic copper when the heat is sufficiently high; heated in the forceps, the outer flame is colored green. With fluxes and Sd it behaves like oxide of copper (v. Table II, 8).

It dissolves in acids with effervescence; also soluble in ammonia.

### *Azurite* [Blue Malachite].

§ 144. $2(CuO.CO^2)+CuO.HO$. H=3.5—4. G=3.5—3.8. Occurs usually crystallized, or in globular masses of columnar structure. It is easily distinguished by its fine blue color; either earthy or vitreous in lustre.

Before the Blp, and to solvents, it behaves like malachite.

### *Copper Vitriol* [Cyanosite].

§ 145. $CuO.SO^3+5HO$. H=2.5. G=2.21. Lustre vitreous; color various shades of blue; taste metallic and nauseous.

Heated in a matrass, swells up, gives out water, and becomes white. On Ch, colors the outer flame green, fuses, and affords a button of metallic copper, crusted with a coat of sulphide. After calcination, gives with fluxes the reactions of copper, sometimes also those of iron.

Soluble in water; a polished plate of iron introduced into the solution becomes coated with copper.

### *Phosphochalcite.*

§ 146. $3CuO.PO^5+3(CuO.HO)$, sometimes $2(3CuO.PO^5)+HO+4(CuO.HO)$. H=4.5—5. G=4—4.4. Occurs both crystallized and massive. Of adamantine lustre, and dark emerald-green or blackish-green color.

In a matrass, gives out water and blackens. A piece, previously heated in a matrass, fuses in the forceps to a black globule, which becomes crystalline on cooling. With Bx and S Ph, behaves like oxide of copper. Strongly heated on Ch with a sufficient quantity of Sd, nearly all the copper is obtained as a metallic globule. Mixed with an equal volume of metallic lead and fused on Ch, a globule of metallic copper is obtained, surrounded by a fused mass of phosphate of lead, which on cooling crystallizes.

Soluble in nitric acid, and in caustic ammonia.

### *Olivenite.*

§ 147. $3CuO. (AsO^5.PO^5)+CuO.HO$. H=3. G=4.1—4.4. Crystallized, or in globular and reniform masses, of indistinctly fibrous structure. Color usually olive-green.

In a matrass, yields a little water. In the forceps, fuses to a globule and colors the outer flame bluish-green; the fused mass crystallizes on cooling. On Ch, fuses with detonation and emission of arsenical vapors to a metallic globule; the globule is white and somewhat brittle, and covered with a brown scoria. Fused with metallic lead, it is decomposed in the same manner as the preceding ore.

Dissolves in nitric acid, also in ammonia.

*Tyrolite.*

§ 148. $[(3CuO.AsO^5+8HO)+2(CuO.HO)]+CaO.CO^2$. H=1—2. G=3. Usually reniform, massive; structure radiate foliaceous. Color pale-green. Very sectile.

Heated in a matrass, decrepitates, yields much water, and blackens. On Ch, fuses with emission of arsenical vapors to a gray scoriaceous mass, in which minute globules of metallic copper occasionally appear. When the mineral is fused on Ch, with addition of Sd and Bx, until the oxide of copper is completely reduced and the slag dissolved in hydrochloric acid, a solution is obtained in which the presence of lime may be shown by the proper reagents.

Dissolves in nitric acid with effervescence, also in ammonia.

*Chrysocolla.*

§ 149. $3CuO.\ 2SiO^3+6HO$. H=2—3. G=2. Occurs usually as an incrustation. It very much resembles malachite; its color is bluish-green, and it is remarkable for its great compactness; its surface is very smooth, giving it the appearance of an enamel or a well-fused slag.

In a matrass, yields water and blackens. In the forceps infusible, coloring the outer flame intensely green. On Ch in O Fl blackens, in R Fl turns red. S Ph and Bx dissolve it with the usual indications of copper; the S Ph bead shows a cloud of undissolved silica. With Sd on Ch, affords globules of metallic copper.

It is decomposed by acids, silica remaining undissolved.

## Ores of Gold, Platinum, and Iridium.

*Native Gold.*

§ 150. Combination of Au and Ag in variable proportions, sometimes with traces of Fe and Cu. H=2.5—3. G=15.6—19.5.

Easily distinguished by its malleability, its cutting like lead, its high specific gravity, and its resistance to acids. Color and streak various shades of gold-yellow. It usually occurs in variously contorted and branched filaments, in scales, in plates, or in small irregular masses.

On Ch, fuses to a globule which, after cooling, has a bright metallic surface. With SPh in OFl, a bead is formed which opalizes on cooling, or becomes opaque and yellow, according to the amount of silver which it contains.

Resists the action of heated concentrated nitric acid; soluble only in aqua regia.

### *Graphic Tellurium* [Sylvanite].

§ 151. $AgTe+2AuTe^3$. H=1.5—2. G=5.7. Of metallic lustre and steel-gray color. Very sextile.

In an open glass tube, yields a white sublimate which, when played upon with the flame, fuses to transparent drops. On Ch, fuses to a dark-gray globule, depositing at the same time a white Ct which, when touched with the RFl, disappears, tinging the flame bluish-green (see §§ 29, 35). It finally affords a light-yellow malleable globule of metallic lustre.

Soluble in aqua regia, leaving a residue of chloride of silver. The solution gives a white precipitate with water.

### *Native Platinum.*

§ 152. Pt, usually combined with a little Fe, Ir, Os, Pd, Rh, and sometimes Cu and Pb.

H=4—4.5. G=16—19. Usually occurs in grains of silver-whitish or gray color, malleable and ductile.

Infusible before the Blp and not acted upon by fluxes. Soluble only in heated aqua regia. The solution gives a yellow granular precipitate with chloride of potassium.

### *Osmium-Iridium* [Iridosmine].

§ 153. The light variety $IrOs^3$ and $IrOs^4$. H=6—7. G=19.3—21.1. Occurs usually in irregular flattened grains, of metallic lustre and tin-white color; but little malleable.

Infusible before the Blp; when fused with nitre in a matrass, the characteristic osmium odor is produced. The fused mass is soluble in water; the solution gives, on addition of nitric acid, a green precipitate. The dark varieties lose before the Blp the metallic lustre, and, when held in the alcohol flame, impart to it a yellowish-red color and great luminating power.

Not visibly affected by any acid.

## Ores of Iron.

### *Meteoric Iron.*

§ 154. Fe with variable quantities of Ni (from 1 to 20 per cent.) and traces of Co, Mg, Mn, Sn, Cu, Cr, Si, C, Cl, S, and P. H=4.5. G=7.3—7.8, rarely as low as 6. Lustre metallic; color iron-gray; ductile; strongly attracted by the magnet.

Infusible. On Ch with Bx or S Ph gives only the reactions of iron. To detect the presence of the other heavy metals, the assay-piece must be dissolved in aqua regia, the liquid mixed with ammonia in excess, filtered, and the ammoniacal filtrates precipitated with sulphydrate of ammonia. The precipitate consists of the sulphides of nickel, cobalt, manganese, and copper, which may be collected on a filter and treated with Bx on Ch as described § 70.

### *Brown Hematite* [Limonite].

§ 155. 2 $Fe^2O^3$. 3HO. H=5—5.5. G=3.6—4. Of a dull brownish-yellow color, earthy or semi-metallic in appearance, and often in mammillary or stalactitic forms.

In a matrass, yields water, and red sesquioxide remains; in platinum forceps, fusible on the edges; gives with Bx and S Ph an iron reaction; the clayey varieties treated with S Ph give a cloud of undissolved silica; treated with Sd and nitre on platinum-foil, the manganese reaction is almost always obtained.

### *Specular Iron* [Hematite].

§ 156. $Fe^2O^3$. H=5.5—6.5. G=4.5—5.3. Of a dark steel-gray or iron-black color, and usually of metallic lustre; its powder is red.

Infusible alone; becomes magnetic after roasting, and gives the usual indications of iron with the fluxes; its powder dissolves readily in heated hydrochloric acid. Contains sometimes chromium and titanium, which may be detected by the processes given in §§ 68 and 111.

### *Magnetic Iron Ore* [Magnetite].

§ 157. $FeO.Fe^2O^3$. H=5.5—6.5. G=4.9—5.2. Its color is iron-black, with a shining metallic or glimmering lustre; its powder is black; it is strongly attracted by the magnet.

It fuses with difficulty, and gives the usual reactions of iron with the fluxes; the pulverized mineral dissolves completely in hydrochloric acid.

### *Iron Pyrites.*

§ 158. $FeS^2$. H=6—6.5. G=4.8—5. Occurs commonly in cubes. Usually of a brass-yellow color and metallic lustre. By its superior hardness, not yielding to the knife, and emitting sparks when struck with steel, it may be distinguished from copper pyrites.

Heated in a glass tube closed at one end, usually emits some sulphuretted hydrogen, and yields a sublimate of sulphur; the residue is attracted by the magnet. Heated on Ch with the O Fl, the sulphur burns off with a blue flame, and leaves red oxide behind, which, when treated with the fluxes, gives pure iron reactions. But slightly affected by hydrochloric acid; nitric acid dissolves it, leaving a residue of sulphur.

### *White Iron Pyrites* [Marcasite].

§ 159. $FeS^2$. H=6—6.5. G=4.6—4.8. Crystals are prismatic. Color usually light bronze-yellow, sometimes inclined to green or gray; occurs frequently in radiated masses or crest-like aggregations. Very liable to decomposition.

Before the Blp it behaves like the preceding.

### *Magnetic Pyrites* [Pyrrhotine].

§ 160. $5FeS+Fe^2S^3$. H=3.5—4.5. G=4.4—4.7. Very much resembles common iron pyrites, from which it is distinguished by its inferior hardness, and by being slightly attracted by the magnet.

Heated in a matrass, remains unchanged; in the open glass tube, emits sulphurous acid but yields no sublimate. On Ch in RFl, fuses to a globule, which is covered with an uneven black coating, which follows the magnet, and which, on a surface of fracture, exhibits a yellowish crystalline structure and metallic lustre. In OFl it is converted into red oxide.

Soluble in hydrochloric acid, excepting the sulphur, with evolution of sulphuretted hydrogen.

### *Arsenical Pyrites* [Mispickel].

§ 161. $FeS^2+FeAs$. H=5.5—6. G=5—6.4. Of metallic lustre and a silver-white color. Streak dark grayish-black. Brittle.

Heated in a matrass, yields first a red sublimate of sulphide of arsenic, and afterwards a black crystalline one of metallic arsenic; in an open glass tube, yields arsenous acid and sulphurous acid. On Ch, emits copious arsenical fumes, and a Ct of arsenous acid is deposited; then fuses to a globule which shows the properties of fused magnetic pyrites. Frequently contains cobalt, the presence of which may be detected by the method described in § 69.

Soluble in nitric acid and aqua regia, leaving a residue of sulphur and arsenous acid; the latter dissolves with continued digestion.

### *Titaniferous Iron* [Ilmenite].

§ 162. $Ti^2O^3$ and $Fe^2O^3$ in various proportions. H=5—6. G=5.5—5. Of iron-black color, usually in tabular crystals, bears a great resemblance to specular iron, but gives no red powder.

Alone in the OFl infusible; in RFl it may be rounded at the edges. With Bx and SPh in OFl, gives the reactions of pure oxide of iron; but the SPh bead when treated with the RFl assumes a brownish-red color, the intensity of which depends upon the amount of titanic acid present; this glass, when treated with tin on Ch, turns violet (v. Table II, 23). To show conclusively the presence of Ti, follow the method given in § 111.

Dissolved by hydrochloric acid and aqua regia with separation of titanic acid; some varieties dissolve with great difficulty, even when reduced to a very fine powder.

### *Spathic Iron* [Chalybite].

§ 163. $FeO.CO^2$. H=3.5—4.5. G=3.7—3.9. Color from grayish-yellow to reddish-brown; crystallizes in rhombohedrons, which are often curved, and are very distinctly cleavable; often massive.

Heated in a matrass, frequently decrepitates, carbonic acid and carbonic oxide are given out, and a black oxide of iron remains, which is attracted by the magnet. Alone, infusible. With Bx and SPh it gives the pure iron reactions, and with Sd sometimes those of manganese. It dissolves in strong acids with effervescence, but with difficulty, and only when pulverized.

*Green Vitriol* [Copperas].

§ 164. $FeO. So^3+7HO$. H=2. G=1.83. Occurs usually massive and pulverulent, of various shades of green, becoming yellowish on exposure to air; taste astringent and metallic.

In a matrass, gives out sulphurous acid and water, which shows acid reaction. Strongly heated, only sesquioxide of iron remains.

Soluble in water.

*Vivianite.*

§ 165. $6(3FeO,Po^5+8HO)+(3Fe^2O^3,2PO^5+8HO)$. H=1.5—2. G=2.66. Occurs crystallized, or in reniform and globular masses, sometimes as incrustation. Color blue to green, usually dirty blue.

In a matrass, swells and gives pure water. In the forceps, fuses to a steel-gray metallic globule, coloring the outer flame bluish-green. With fluxes gives the reactions of iron.

Easily soluble in hydrochloric acid and nitric acid. With a solution of caustic potassa, it blackens.

*Scorodite.*

§ 166. $Fe^2O^3,AsO^5+4HO$. H=3.5—4. G=3.1—3.3. Crystallized. Color pale leek-green or liver-brown.

In a matrass, yields pure water. In the forceps, fuses to a gray scoriaceous slag of metallic lustre, coloring the outer flame pale-blue. On Ch, emits arsenical vapors and fuses to a gray magnetic slag, of metallic lustre, which gives with fluxes the reactions of iron.

Not affected by nitric acid; forms a brown solution with hydrochloric acid; partially dissolved by ammonia, leaving a brown residue.

### Ores of Lead.

*Plumbic Ochre.*

§ 167. PbO, containing frequently $PbO.CO^2$, CaO, $Fe^2O^3$, and $SiO^3$. G=8. Massive. Lustre dull; color between sulphur and orpiment-yellow.

Before the Blp, behaves like oxide of lead.

*Minium.*

§ 168. $PbO,Pb^2O^3$. G=4.6. Pulverulent. Color vivid red, mixed with yellow.

Before the Blp, behaves like oxide of lead.

With hydrochloric acid, evolves chlorine and is converted into chloride of lead. With nitric acid, becomes brown.

*Galena.*

§ 169. PbS. H=2.5—2.75. G=7.25—7.7. Color, lead-gray; of metallic lustre. Crystals usually affect the cubical form, and possess very perfect cubic cleavage.

Heated in a matrass, sometimes decrepitates and frequently yields a slight white sublimate. Heated in an open glass tube, emits sulphurous acid, and, on the heat being raised, gives a white sublimate of sulphate of lead. Heated on Ch, affords a globule of pure lead, the Ch becoming at the same time covered with sulphate of lead and oxide of lead. The globule of metallic lead yields generally a little silver on cupellation. The presence of antimony is ascertained as shown § 49. Zinc, § 113. Iron, § 83.

It dissolves with some difficulty in boiling hydrochloric acid, with evolution of sulphuretted hydrogen. Very dilute nitric acid has no effect on it, but by a stronger acid it is readily dissolved with evolution of nitrous acid vapors. By fuming nitric acid and aqua regia it is very violently acted upon, being converted into sulphate, or a mixture of the sulphate with the chloride.

*Bournonite.*

§ 170. $3Cu^2S,SbS^3+2(3PbS,SbS^3)$. H=2.5—3. G=5.7—5.9. Occurs crystallized, and massive, granular, compact; lustre metallic; color and streak steel-gray.

In a matrass, decrepitates and yields with a strong heat a dark-red sublimate. In an open tube, sulphurous acid is evolved and abundant antimonial fumes, which condense partly on the upper and partly on the lower side of the tube; the former consist of antimonious acid, which is volatile; the latter is not volatile, and consists of a mixture of antimonate of oxide of antimony with antimonate of lead. On Ch, fuses readily to a black globule and deposits a Ct of antimonious acid; with strong heat a Ct of oxide of lead is obtained; the remaining globule, when treated with Bx in OFl, gives the reactions of copper, and the globule assumes the appearance of metallic copper.

Dissolves readily in nitric acid to a blue liquid, leaving a residue of antimonious acid and sulphur. Aqua regia leaves a residue of sulphur, chloride of lead, and antimonite of lead; the solution gives a precipitate with water. Ammonia dissolves a portion of the sulphide of antimony.

The following ores behave before the Blp in a very similar manner.

*Geocronite.* PbS, $(SbS^3, AsS^3)+4PbS$.

*Dufrenoysite.* PbS, $AsS^3+Pbs$.

*Boulangerite.* PbS, $SbS^3+2PbS$.

*Heteromorphite.* PbS, $SbS^3+PbS$.

*Jamesonite.* $2(PbS, SbS^3)+PbS$.

*Plagionite.* $3(PbS, SbS^3)+PbS$.

*Zinkenite.* PbS, $SbS^3$.

Those minerals in which a part of the $SbS^3$ is substituted by $AsS^3$, give on Ch arsenical vapors, and in an open tube a crystalline sublimate.

*Cerasine* [Corneous Lead].

§ 171. $PbCl+PbO.CO^2$. H=2.75—3. G=6—6.3. Forms crystals of adamantine lustre, of white, gray, or yellow color.

In a matrass, decrepitates slightly and becomes a little darker yellow. On Ch, fuses readily, emits acid vapors, becomes reduced to metallic lead, and gives a white Ct of chloride of lead and a yellow Ct of oxide.

Dissolves in nitric acid with effervescence.

*White Lead Ore* [Cerusite].

§ 172. PbO. $CO^2$. H=3—3.5. G=6.4. Occurs granularly massive, or in prismatic needles, or compressed plates. Color mostly white, yellow, or gray.

When heated in a matrass, decrepitates and turns yellow; carbonic acid is given out. Heated on Ch alone, is reduced to metallic lead. Treated with fluxes, dissolves with effervescence and gives the reactions of pure oxide of lead (v. Table II, 12); dissolves readily and with effervescence in dilute nitric acid; with hydrochloric acid, leaves a residue of chloride of lead; dissolves in a solution of caustic potassa.

F

*Leadhillite.*

§ 173. $PbO.SO^3+3(PbO.CO^2)$. H=2.5. G=6.2—6.5. Occurs in transparent crystals of pearly or resinous lustre. Color white, passing into yellow, green, or gray.

On Ch, intumesces slightly, becomes yellow, but white again on cooling; with greater heat easily reduced to metallic lead.

Dissolves in nitric acid with effervescence, leaving a residue of sulphate of lead.

*Lead Vitriol* [Anglesite].

§ 174. $PbO.SO^3$. H=2.75—3. G=6.2. It often occurs in small octahedral crystals with many facets, but more frequently in laminar masses; of high lustre.

Heated in a matrass, decrepitates and usually yields a little water. Treated on Ch in OFl, fuses to a clear bead, which on cooling turns milk-white; with Sd on Ch, affords a globule of metallic lead; the Sd is absorbed by the Ch and shows, when placed on silver-foil, a strong sulphur reaction. With the fluxes, gives the reactions of oxide of lead. Traces of iron or manganese may be detected by Bx or Sd as shown §§ 83 and 90.

It dissolves in acids only with great difficulty; by hydrochloric acid it is partly decomposed; the pulverized mineral is soluble in a solution of caustic potassa.

*Phosphate of Lead* [Pyromorphite].

§ 175. Essentially $PbCl+3(3PbO.[PO^5.AsO^5])$. H=3.5—4. G=6.5—7. It occurs often in globular masses with a columnar structure, also fibrous and granular. Color green, yellow, and brown.

Heated in a matrass, sometimes decrepitates and yields, with continued heat, a faint white and volatile sublimate of chloride of lead. Heated in the platinum-pointed pincers, fuses readily and colors the outer flame blue; if the amount of phosphoric acid is not too small, the edges of the flame will appear green. With SPh and oxide of copper, gives the reaction for chlorine, § 65. On Ch in the O Fl, fuses to a globule, which on cooling assumes a polyhedral form and a dark color; in the R Fl, yields a Ct of oxide of lead, and the globule, on cooling, assumes dodecahedral facets of pearly lustre. With boracic acid and iron wire, gives the reaction for phosphoric acid (§ 95). With Sd on Ch, affords metallic lead.

Some varieties contain arsenic acid, which is readily detected by the odor when treated with Sd on Ch (§ 54).

Soluble in nitric acid, and solution of caustic potassa.

*Plumbo-Resinite.*

§ 176. $3PbO.\ PO^5+6(Al^2O^3,3HO)$. H=4—4.5. G=6.3—6.4. In reniform or globular masses, with a columnar structure; also compact massive. Of resinous lustre; color usually yellowish-brown; resembling gum-arabic in appearance.

In a matrass, decrepitates and gives out water. In the forceps, intumesces and colors the outer flame azure-blue. On Ch, intumesces, becomes white and opaque, and fuses but imperfectly, depositing a faint white Ct of chloride of lead. In small quantities, soluble in Bx and SPh to clear beads. With Sd on Ch, minute globules of metallic lead are obtained. Treated with SoCo, assumes a fine blue color.

Soluble in nitric acid.

*Red Lead Ore* [Crocoisite].

§ 177. $PbO.\ CrO^3$. H=2.5—3. G=5.9—6.1. Occurs usually in bright hyacinth-red crystals of adamantine lustre.

In a matrass, decrepitates; the crystals are broken up into minute pieces and assume a darker color. On Ch, fuses and becomes reduced with detonation; a Ct of oxide of lead is formed, and grayish-green sesquioxide of chromium remains with the metallic globule. With Sd on Ch, affords a globule of metallic lead. With Sd on platinum foil, fuses to a dark-yellow mass, which becomes green in RFl. With Bx or SPh in OFl, dissolved; the bead appears yellow while hot, but becomes green on cooling. Fused in a platinum spoon with from 3 to 4 parts of bisulphate of potassa, gives a dark-violet mass, which is greenish-white when cold.

Dissolves in heated hydrochloric acid to a green liquid, leaving a residue of chloride of lead. Dissolves with difficulty in nitric acid to a yellowish-red liquid. A solution of caustic potassa colors it brown, and finally dissolves it to a yellow liquid.

*Vauquelinite.*

§ 178. $3CuO,2CrO^3+2(3PbO,2CrO^3)$. H=2.5—3. G=5.5—5.7. Occurs usually in minute crystals, or in reniform or granular masses. Color dark-green to brown, sometimes nearly black.

On Ch, fuses with effervescence to a gray submetallic globule; where the mass is in contact with the coal, small globules of lead make their appearance; in RFl a Ct of oxide of lead is formed. With Bx or SPh in O Fl, clear green beads are obtained, which remain green on cooling, but which on application of the RFl become red and opaque; this reaction appears most distinctly on Ch with Sn. With Sd on platinum wire in OFl, dissolves to a transparent green bead, which on cooling becomes yellow and opaque; on treating the bead with a few drops of water, a yellow solution is obtained, in which the presence of chromic acid may be proved as described § 68. With Sd on Ch, is completely decomposed; on treating the reduced metals with boracic acid on Ch (v. § 71) a globule of metallic copper is obtained.

Partly soluble in nitric acid to a dark green liquid; the residue is yellow.

### *Wulfenite* [Yellow Lead Ore].

§ 179. PbO, $MoO^3$, sometimes with a little $CrO^3$. H=2.75—3. G=6.3—6.9. Crystallized or granularly massive, firmly coherent. Color usually wax-yellow, passing into orange-yellow.

In a matrass, decrepitates and becomes darker while hot. On Ch, fuses and is partly absorbed by the coal, while metallic lead and a Ct of oxide of lead are deposited. With Bx or SPh on platinum wire gives the reactions of molybdic acid (v. Table II, 15). With Sd on Ch, affords a globule of metallic lead. Fused with bisulphate of potassa in a platinum spoon, a yellowish mass is obtained, which becomes white on cooling; treated with distilled water and a piece of metallic zinc placed into the solution, the liquid assumes a blue color.

Dissolves in concentrate hydrochloric acid to a green liquid, leaving a residue of chloride of lead. The pulverized mineral is decomposed on being digested with nitric acid; a yellowish-white residue is left, which becomes blue when exposed to air in thin layers.

## Ores of Manganese.

### *Pyrolusite* [Gray Ore of Manganese].

§ 180. $MnO^2$. H=2—2.5. G=4.8. Of black or dark-gray color and little lustre; powder black; sometimes of columnar structure.

In a matrass, usually yields a little water; when heated to redness, oxygen is evolved. Alone infusible, but turning reddish-brown when the temperature is sufficiently high. Soluble in Bx and SPh with the usual manganese-reactions; gives frequently the indications of iron.

Soluble in hydrochloric acid with disengagement of chlorine.

*Hausmannite* [Black Manganese].

§ 181. $MnO, Mn^2O^3$. H=5—5.5. G=4.7. Crystallized, or granular, particles strongly coherent. Color brownish-black; streak chestnut-brown.

Before the Blp, and to hydrochloric acid behaves like the preceding ore.

*Braunite.*

§ 182. $Mn^2O^3$. H=6—6.5. G=4.7—4.8. Occurs crystallized or massive. Color and streak dark brownish-black.

In a matrass, does not give any water; behaves otherwise like pyrolusite. Dissolves in hydrochloric acid with disengagement of chlorine, leaving sometimes a residue of silica.

*Psilomelane.*

§ 183. Composition very various, essentially $Mn^2O^3$ with BaO or KO, and HO. H=5—6. G=3.7—4.3. Massive. Color iron-black; streak brownish-black, shining.

Before the Blp and to solvents it behaves like pyrolusite.

*Wad* [Bog Manganese].

§ 184. Essentially $MnO^2$, MnO, and HO; contains often $Fe^2O^3$, $Al^2O^3$, BaO, $SiO^3$, &c. H=0.5—6. G=3—4.2. Amorphous, earthy or compact, of a dull black color.

In a matrass, yields water abundantly, and otherwise behaves like pyrolusite. Some varieties, known under the name of "Cupreous Manganese," when treated with Sd and Bx on Ch, afford a globule of metallic copper.

*Diallogite.*

§ 185. $MnO,CO^2$ when pure, sometimes (MnO, FeO, CuO, MgO), $CO^2$. H=3.5—4.5. G=3.4—3.6. Occurs crystallized, or in globular masses of columnar structure; also massive. Color shades of rose-red, brownish; streak white.

In a matrass, some varieties give a little water and decrepitate violently. Infusible. Some varieties, when heated in R Fl, become magnetic. Dissolves in fluxes with effervescence and gives usually the reaction of manganese and iron.

The pulverized mineral is little affected by hydrochloric acid in the cold; on heating dissolves with effervescence.

*Franklinite.*

§ 186. ZnO, $Mn^2O^3+4\ Fe^2O^3$. H=5.5—6.5. G=5. Occurs crystallized, and massive. Lustre metallic; color iron-black; streak dark reddish-brown; acts slightly on the magnet.

Infusible. Dissolves in Bx and S Ph with manganese-reaction; the Bx bead, when treated on Ch in R Fl, becomes bottle-green. With Sd on platinum foil, gives manganese-reaction. With Sd on Ch, gives a faint Ct of oxide of zinc, which becomes more distinct on addition of Bx.

Dissolves completely in heated hydrochloric acid to a greenish-yellow liquid, chlorine being evolved.

## Ores of Mercury.

*Native Mercury.*

§ 187. Hg, sometimes containing a little Ag. G=13.5 Metallic globules of a tin-white color.

Heated in a matrass, is converted into vapor, which condenses in the neck of the matrass to small metallic globules.

Dissolves readily in nitric acid.

*Amalgam.*

§ 188. $AgHg^2$ and $AgHg^3$. H=3—3.5. G=10.5—14. Occurs crystallized and massive. Color and streak silver-white; opaque.

In a matrass, boils, gives a sublimate of metallic mercury, and leaves a spongy residue of silver, which on Ch fuses readily to a globule.

Dissolves readily in nitric acid.

*Calomel* [Horn Quicksilver].

§ 189. $Hg^2Cl$. H=1—2. G=6.48. Occurs usually in distinct crystals or crystalline coats, of adamantine lustre and yellowish-gray color.

In a matrass, yields a white sublimate of subchloride of mercury. Mixed with Sd and heated in a matrass, affords globules of metallic mercury. On Ch, completely volatilized, giving a white Ct. Shows the chlorine-reaction when treated as described § 65.

Treated with boiling hydrochloric acid, is partly dissolved and becomes gray. Not affected by nitric acid, dissolved by aqua regia. With a solution of caustic potassa, becomes black.

*Cinnabar.*

§ 190. HgS. H=2—2.5. G=8.9. Color various shades of red, from cochineal-red to dark brownish-red. Powder always bright-red. It occurs in very small flattened crystals, or granularly massive.

Heated in a matrass, is volatilized and condenses to a black sublimate, which by friction assumes a red color. Mixed with Sd, yields on heating globules of metallic mercury. In an open glass tube, is partially decomposed into metallic mercury and sulphurous acid. On Ch it is, when pure, wholly volatilized.

Nitric acid and hydrochloric acid have no visible effect on it. Aqua regia dissolves it, part of the sulphur being precipitated. Insoluble in caustic potassa.

## Ores of Nickel.

*Copper Nickel.*

§ 191. $Ni^2$ As or $Ni^2$ (As. Sb). H=5—5.5. G=7.3—7.6. Usually massive; of copper-red color, with a gray tarnish, and metallic lustre; very brittle.

In a matrass, affords a very slight sublimate of arsenous acid. In an open glass tube, yields a copious sublimate of arsenous acid, and usually a little sulphurous acid; the assay-piece assumes at the same time a yellowish-green color and crumbles to powder. On Ch, emits arsenical fumes and fuses to a white and brittle globule which, when treated with Bx in RFl, imparts usually to the flux the colors of iron and cobalt. Sometimes a faint Ct of oxide of lead is deposited on the Ch.

Dissolves almost completely in concentrated nitric acid; the solution has a green color; on cooling arsenous acid separates. Readily dissolved by aqua regia.

### *Nickel Glance* [Gersdorffite].

§ 192. $(Ni,Fe)+(S^2,As)$. H=5.5. G=5.6—6.9. Of silver-white or steel-gray color, and metallic lustre.

In a matrass, decrepitates violently and yields a yellowish-brown sublimate of sulphide of arsenic. In an open glass tube, emits arsenous acid and sulphurous acid. On Ch, fuses with emission of sulphur and arsenical fumes to a globule which, when treated with Bx in RFl, gives the reäctions of iron and cobalt. After having removed these two metals, the remaining globule exhibits with the fluxes the reactions of pure oxide of nickel.

Partly dissolved by nitric acid, sulphur and arsenous acid being precipitated.

### *Nickeliferous Gray Antimony* [Ullmannite].

§ 193. $NiS^2+Ni(Sb,As)$. H=5—5.5. G=6.2—6.5. It closely resembles the preceding ore in its physical properties.

In a matrass, yields a slight white sublimate. In an open glass tube, emits copious antimonial fumes and sulphurous acid. On Ch in RFl, fuses to a globule, and coats the Ch with antimonious acid; sometimes the odor of arsenic is observable. The melting globule, when treated with Bx, frequently exhibits the reactions of iron and cobalt besides those of nickel.

It is violently acted upon by concentrated nitric acid, sulphur, antimonious and arsenous acids being precipitated. Aqua regia dissolves it, excepting the sulphur, to a green liquid.

### *Capillary Pyrites* [Millerite].

§ 194. NiS. H=3—3.5. G=5.2—5.6. Occurs usually in delicate capillary crystals of brass-yellow color and metallic lustre.

In an open glass tube, evolves sulphurous acid. On Ch, fuses with emission of sparks to a metallic globule which is attracted by the magnet. The calcined mineral gives with fluxes the indications of oxide of nickel, and sometimes also those of oxide of cobalt.

By heated concentrated nitric acid it is but little affected, but its color is changed to gray. By aqua regia it is wholly dissolved.

### *Emerald Nickel.*

§195. $(NiO.CO^2+4HO)+2(NiO.HO)$. H=3—3.2. G=2.5—2.7. Usually forms incrustations of emerald-green color, and vitreous lustre.

In a matrass, loses already at 212° a considerable amount of water, and blackens. In Bx and SPh, dissolves with effervescence, exhibiting the characteristic nickel-reactions.

Dissolves easilv in heated dilute hydrochloric acid with effervescence.

### *Annabergite* [Nickel Green].

§ 196. $3NiO.AsO^5+8HO$. Soft. In capillary crystals, also massive and disseminated. Color fine apple-green.

In a matrass, yields water and darkens in color. In the forceps, fuses and colors the outer flame light-blue. On Ch in RFl, fuses with emission of arsenical vapor to a blackish-gray globule; when treated with Bx the globule gives the reactions of nickel, sometimes also those of iron and cobalt.

Soluble in acids.

## ORES OF SILVER.

### *Native Silver.*

§ 197. Pure silver, associated with gold, copper, arsenic, iron, and other metals. H=2.5—3. G=10—11. Color silver-white; lustre metallic; ductile and malleable. Occurs usually in twisted filaments, or arborescent; sometimes in plates or massive.

On Ch, fuses easily to a globule, which assumes a bright surface, and shows after cooling a silver-white color. Foreign metals are detected by the methods given §§ 103–105.

It dissolves nitric acid.

### *Antimonial Silver* [Discrasite].

§ 198. $Ag^6Sb$ and $Ag^4Sb$. H=3.5—4. G=9.4—9.8. Occurs crystallized or massive, granular. Lustre metallic; color and streak silver-white.

On Ch, fuses readily to a gray non-ductile globule and coats the Ch with oxide of antimony; with continued heat the globule assumes the appearance of pure silver, and the Ct becomes reddish.

Dissolves in nitric acid, leaving a residue of oxide of antimony.

### *Horn Silver* [Kerargyrite].

§ 199. Ag Cl. H=1—1.5. G=5.5. Remarkable for its pearl-gray or greenish color, its semi-transparency, resinous lustre, and more especially for its softness, which is so great as to allow it to be marked by the nail. It turns brown on exposure to air. When rubbed with a moistened plate of zinc or iron the latter becomes covered with a coating of silver.

It fuses in a candle-flame. On Ch, is easily reduced, especially when mixed with Sd. Mixed with oxide of copper and heated on Ch in RFl, chloride of copper is formed, which colors the flame azure-blue (v. § 65).

Insoluble in water and nitric acid. Slowly soluble in caustic ammonia. Partially decomposed by a boiling solution of caustic potassa.

### *Embolite* [Chloro-bromide of Silver.]

§ 200. 2AgBr+3AgCl. H=1—1.5. G=5.3—5.4. Crystallized or massive. Lustre resinous; color various shades of green.

On Ch, fuses readily, evolves pungent vapors of bromine, and affords a globule of metallic silver. With Sd on Ch, reduced; on dissolving in water the alkaline mass which has passed into the coal, evaporating the solution to dryness, and treating the residue with bisulphate of potassa as described § 63, bromine-vapors are given out. Fused with oxide of copper on Ch in RFl, colors the outer flame greenish, then blue (v. § 65).

### *Bromyrite* [Bromic Silver].

§ 201. Ag Br. H=1—2. G=5.8—6. Occurs usually in small concretions. Lustre splendent; color yellowish-green or green. Sectile.

Its action before the Blp not known; behaves probably like the preceding.

Only slightly affected by acids. Dissolves in heated concentrated ammonia.

### *Iodyrite* [Iodic Silver].

§ 202. AgI. Soft. G=5.5. Occurs crystallized or in thin plates with a lamellar structure. Color citron-yellow to yellowish-green.

On Ch, fuses readily, colors the flame purple-red, and affords a globule of silver.

### *Silver Glance.*

§ 203. AgS. H=2—2.5. G=7. Color blackish lead-gray; lustre

metallic. It is easily distinguished from other minerals of the same color by being cut by a knife like lead.

On Ch in OFl, intumesces, gives out sulphurous acid, and finally yields a globule of metallic silver.

Soluble in dilute nitric acid, leaving a residue of sulphur.

### *Ruby Silver* [Pyrargyrite. Dark-red Silver Ore].

§ 204. $3AgS, SbS^3$. H=2—2.5. G=5.7—5.9. Color dark-red to black, giving a cochineal-red powder. Crystallizes in hexagonal prisms.

In a matrass, fuses very readily and yields with continued heat a sublimate of tersulphide of antimony. In an open glass tube gives antimonial fumes and sulphurous acid. On Ch, fuses readily and deposits a Ct of antimonious acid, being converted into sulphide of silver; if for a long time exposed to the OFl or, when mixed with Sd, to the RFl, affords a globule of metallic silver.

Part of the $SbS^3$ is sometimes substituted by $AsS^3$; it then gives out arsenical fumes when mixed with Sd and heated in the R Fl on Ch.

The pulverized mineral, when heated with nitric acid, turns black and is ultimately dissolved, leaving a residue of sulphur and antimonious acid. Caustic potassa also blackens it and effects partial solution, from which acids precipitates tersulphide of antimony.

### *Proustite* [Light-red Silver Ore].

§ 205. $3AgS, AsS^3$. H=2—2.5. G=5.4—5.5. Very much resembles the dark-red silver ore, but is of a somewhat lighter color.

Before the Blp and to solvents, behaves like the preceding, excepting it gives off arsenical fumes instead of antimonious acid. The solution in caustic potassa deposits a yellow precipitate when neutralized with acids.

### *Brittle Silver Ore* [Stephanite].

§ 206. 6 $AgS, SbS^3$. H=2—2.5. G=6.2. Of metallic lustre and iron-black color; it is very brittle and fragile, and its powder black.

In a matrass, decrepitates, then fuses and ultimately yields a faint sublimate of tersulphide of antimony. On Ch, fuses very readily and coats the Ch with antimonious acid. If the blast with the O Fl is kept up for a sufficient time, the Ct assumes a red color and a globule of metallic silver is obtained. Contains frequently copper and iron, which may be detected by the process described § 71. If arsenic is present it gives in the open tube a crystalline sublimate of arsenous acid.

In dilute heated nitric acid it dissolves, excepting the sulphur and antimonious acid; the solution becomes milky on addition of water. Partially dissolved by a boiling solution of caustic potassa.

### *Polybasite.*

§ 207. 9 ($Cu^2S$, AgS) ($SbS^3$, $AsS^3$). H=2—3. G=6.2. Occurs usually in short tabular prisms, or massive. Lustre metallic; color and streak iron-black.

In a matrass, fuses very readily, but gives nothing volatile. In an open tube, gives sulphurous acid and antimonial fumes; the sublimate contains sometimes crystals of arsenous acid. On Ch, gives a Ct of oxide of antimony; with continued heat, gives a bright metallic globule, which, on cooling, becomes black on its surface; sometimes a faint Ct of oxide of zinc is deposited; the metallic globule affords with fluxes the reaction of silver and copper.

With acids behaves like bournonite.

### *Stromeyerite* [Argentiferous Sulphide of Copper].

§ 208. $Cu^2S$. + AgS. H=2.5—3. G=6.2—6.3. Occurs usually in small compact masses. Lustre metallic; color dark steel-gray.

In a matrass, fuses easily and gives sometimes a little sulphur. In an open tube, fuses to a globule and gives sulphurous acid. On Ch, fuses to a gray metallic globule which is a little malleable; with fluxes the globule gives the reactions of copper, sometimes also those of iron; on a cupel with lead affords a globule of silver.

Dissolves in nitric acid, leaving a residue of sulphur.

## ORES OF TIN.

### *Tin Ore* [Cassiterite].

§ 209. $SnO^2$. H=6—7. G=6.3—7.1. It occurs crystallized in square prisms terminated by more or less complicated pyramids; reëntrant angles are so frequent that they are to a certain extent characteristic; also massive, and in small mammillated masses of fibrous texture, hence called "wood tin." Color very various, but usually brown or black. The crystals commonly possess a very brilliant lustre.

Infusible in the forceps; the behavior before the Blp is that of pure oxide of tin (v. Table II, 22), excepting of its sometimes imparting to the Bx bead a slight yellowish tinge, owing to the presence of iron, and exhibiting the reaction for manganese when fused with soda and nitre on platinum-foil.

Insoluble in acids. Fused with caustic potassa, yields a mass which is mostly soluble in water.

*Tin Pyrites.*

§ 210. 2 $Cu^2S$, $SnS^2$ + 2 (FeS, ZnS), $SnS^2$. H=4. G=4.3—4.5. Of steel-gray or iron-black color, and metallic lustre. Occurs usually massive, granular, and disseminated.

In an open glass tube, yields sulphurous acid and oxide of tin, which collects close to the assay-piece and which cannot be volatilized by heat. On Ch in R Fl, fuses to a black scoriaceous globule; in O Fl, gives out sulphurous acid and becomes covered with oxide of tin. When well calcined by the alternate application of O Fl and R Fl, gives with Bx the indications of Fe and Cu. With Sd and Bx, yields a globule of impure copper.

Decomposed by nitric acid; a blue solution is obtained, and a mixture of sulphur and oxide of tin remains undissolved.

## Ores of Zinc.

*Red Zinc Ore* [Zincite].

§ 211. ZnO, containing some $Mn^2O^3$. H=4—4.5. G=5.4—5.5. Of a deep-red color and high lustre; of distinctly foliated structure.

Infusible alone. Dissolved by Bx in O Fl with manganese reaction. With Sd on Ch, deposits a copious Ct of oxide of zinc.

Soluble in nitric acid without effervescence; in hydrochloric acid with evolution of chlorine.

*Blende.*

§ 212. ZnS. H=3.5—4. G=3.9—4.2. Of very variable color, from yellow to black; of resinous lustre and lamellar aspect, distinctly cleavable. It occurs often crystallized in rhomboidal dodecahedrons. The powder is always light-colored, white or grayish, and dull.

In a matrass, sometimes decrepitates violently, but gives nothing volatile; its color also remains unchanged, excepting the green varieties, which become yellow. Strongly heated in an open glass tube, sulphurous acid is evolved, and the color of the calcined assay is white, yellowish, or brownish, according to the amount of FeS which it contains. Alone, infusible or only rounded at the thinnest edges. On Ch in R Fl a feeble dark Ct of oxide of cadmium is usually obtained, which is soon followed by a pure zinc-Ct. With Sd on Ch, is easily reduced, and the characteristic zinc-flame may frequently be observed. Iron is readily detected by calcining the mineral in the O Fl and treating the residue with Bx.

The pulverized mineral dissolves in nitric acid, leaving a residue of sulphur.

### *Smithsonite* [Calamine].

§ 213. $ZnO.CO^2$. H=5. G=4—4.5. It is found crystallized in forms derived from the rhomboid. Of vitreous lustre, and white, grayish, or brownish color; semi-transparent or opaque. Often stalactitic or mammillary.

Heated in a matrass, loses carbonic acid and, if pure, appears after cooling enamel-white. The ZnO is often to a large extent substituted by FeO, MnO, CdO, PbO, MgO, CaO; it then, after cooling, frequently assumes a dark color and gives with fluxes the indications of iron and manganese. Mixed with Sd and exposed to the R Fl, it is decomposed and oxide of zinc deposited on the Ch. If the temperature was raised sufficiently high, a zinc-flame is sometimes observable. The Ct is at first dark yellow, or reddish when cadmium is present.

It readily dissolves in acids with effervescence; also in caustic potassa.

### *Calamine.*

§ 214. $3ZnO+SiO^3+2HO$ or $2(3ZnO.SiO^3)+3HO$. H=4.5—5. G=3.1—3.9. It closely resembles in its physical characters the preceding ore. It is electric by heat; the smallest fragment heated attracts light substances.

Infusible in the forceps. In a matrass yields water and turns milk-white. Bx dissolves it to a transparent glass, which cannot be made opaque by flaming. It dissolves in S Ph to a transparent glass which becomes opaque on cooling, and in which, when highly saturated, clouds of silica are observable while hot. With Sd on Ch, swells and affords with difficulty a Ct of oxide of zinc. With SoCo, assumes a green color, which, when the heat is raised, passes into a fine light-blue on the fused edges.

It is readily decomposed by acids, with separation of gelatinous silicic acid. Partly dissolved by caustic potassa.

---

# APPENDIX.

## Fossil Fuel.

### *Anthracite.*

§ 215. C, with a small percentage of $SiO^3$, $Al^2O^3$, and $Fe^2O^3$. H=2—2.5. G=1.3—1.8. Lustre bright, often sub-metallic; color iron-black, frequently iridescent. Fracture conchoidal.

In a matrass, gives usually a little water, but no empyreumatic oil. Heated on platinum foil in O Fl, is slowly consumed without flame, leaving a small quantity of ash, which consists of $SiO^3$, $Al^2O^3$, and more or less $Fe^2O^3$. Does not color a boiling solution of caustic potassa.

### *Bituminous Coal* [Common Coal].

§ 216. C, H, O, in variable proportions; the bituminous matter contains from 76 to 90 per cent. of carbon; the earthy impurities consist principally of $SiO^3$, $Al^2O^3$, and CaO; contains frequently a small amount of N and $FeS^2$. Softer than anthracite, G=1.2—1.5. Less highly lustrous than the preceding, and of a more purely black or brownish-black color.

In a matrass, some varieties soften and cake (*caking coal*), while others are entirely infusible; all varieties are decomposed, evolve combustible gases and empyreumatic oils, and leave a residue of more or less metallic lustre (coke), which behaves like anthracite. On platinum foil, burns with a luminous flame and emission of smoke, leaving an earthy residue.

Boiled with a solution of caustic potassa, or with ether, imparts to these solvents no, or only a pale-yellow, color.

*Brown Coal.*

§ 217. Composition the same as that of bituminous coal, but the organic constituents contain only from 60 to 75 per cent. of carbon. In physical proportion bears sometimes a close resemblance to the preceding; some varieties show distinctly the texture of wood (*lignite*).

In a matrass, infusible, but some varieties soften; evolves combustible gases, empyreumatic oils, water of acid reaction, and a peculiar disagreeable odor, leaving a residue which consists of carbon and a considerable amount of ash. On platinum foil, burns with a smoky flame and emission of a peculiar odor.

Boiled with a solution of caustic potassa, colors the liquid brown.

*Asphaltum.*

§ 218. C, H, O, in variable proportions, with about 75 per cent. of carbon. G=1—1.2. Of black or brownish-black color, and bituminous odor.

Fuses at about 100° C, and burns with a bright flame and emission of a thick smoke, leaving little ash, which consists essentially of $SiO^3$, $Al^2O^3$, and $Fe^2O^3$. In a matrass, gives empyreumatic oil, some ammoniacal water, combustible gases, and leaves a carbonaceous residue.

Treated with boiling ether, colors the solvent wine-red to brownish-red (distinction from bituminous coal); treated with a boiling solution of caustic potassa, does not color the liquid, or imparts at the most a pale-yellow color (distinction from brown-coal).

# FIFTH CHAPTER

## SYSTEMATIC METHOD FOR THE DISCRIMINATION OF INORGANIC COMPOUNDS.

THE careful observer, having become well acquainted with the reactions which are exhibited by the metallic oxides and other simple compounds, when subjected to the various treatments detailed in the second chapter, will find no difficulty in ascertaining the nature of any mineral substance presented to him for analysis.

If the reactions are not quite distinct, owing to an intermixture with other substances, he may call to his aid the processes laid down in the third chapter, which will enable him, in most cases, to detect also the nature of the impurities. But in order to attain satisfactory results in this way, a certain familiarity with all the principal tests is a necessary condition; this once acquired, any further directions are quite superfluous.

Those, however, who have not devoted much time to blow-pipe operations, will sometimes experience some difficulties in drawing the correct conclusions from the observed phenomena, a difficulty which is to a great extent obviated by pursuing the course given below. This methodical course has the advantage of giving the operator the answer to every phenomenon which he observes, and thus leading him, though sometimes by a very tortuous path, to the right solution. An example will show this more clearly, and teach at the same time the use of the table.

Suppose a substance be given for analysis. The operator commences with No. 1. The substance is heated in R Fl on Ch: a garlic odor is disengaged; proceed to No. 2. Treated with Sd on Ch does *not* give a mass which exhibits the reaction of sulphur; proceed to No. 3. The substance shows no metallic aspect; proceed to No. 131, thence to No. 135. It is not wholly volatilized, nor does it exhibit the reaction of sulphur; proceed to No. 137.

Here we find that the substance must either be an arsenite or an arsenate (which of the two, cannot be decided by the Blp alone), and to find the metal, we proceed to No. 102. It affords, after calcination, with Sd on Ch a fusible metallic button; proceed to No. 103. The button is oxidable (because on being heated in OFl, becomes covered with a black coating of oxide); proceed to No. 105. The button is red and malleable; the metal is copper. The substance, therefore, was arsenite, or arsenate, of copper.

The chief constituents of the body having thus been ascertained, the analyst should never omit to test the correctness of the result by the processes laid down in the third chapter. In the example given above, he should verify the result by the test given in § 57 for arsenous acid, and by those given in §§ 71 and 74 for copper. If we wish to examine the assay also for the presence or absence of some accessory constituents, we must always have recourse to the methods detailed in Chapter III. For example, having found the body under trial to consist essentially of sulphur and lead, and it appears desirable to know, whether or not it contains any silver, we must subject it to the treatment described § 103.

| | | |
|---|---|---|
| 1 | On Ch (RFl) with or without Sd disengages a garlic odor | 2 |
| | Not | 4 |
| 2 | With Sd (RFl) on Ch yields a scoriaceous mass which exhibits the sulphur-reaction (§ 107)...*Sulpharsenide.* | 131 |
| | Not | 3 |
| 3 | Metallic aspect...*Arsenide.* | 131 |
| | Not...*Arsenite and Arsenate.* | 131 |
| 4 | On Ch (OFl) disengages sulphurous acid, and exhibits the sulphur-reaction (§ 107)...*Sulphur Compound.* | 125 |
| | Not | 5 |
| 5 | On Ch disengages the odor of rotten horse-radish...*Selenium Compound.* | 136 |
| | Not | 6 |
| 6 | The substance, after having been well dried, fuses on red-hot Ch | 7 |
| | Not | 11 |

| | | |
|---|---|---|
| 7 | Treated as indicated § 65 imparts to the flame an azure-blue or green color | 8 |
| | Not ... *Nitrate.* | 102 |
| 8 | The color is azure-blue | 9 |
| | The color is green | 10 |
| 9 | Treated as indicated § 63 disengages deep yellow vapors. ... *Bromate.* | 102 |
| | Not ... *Chlorate.* | 102 |
| 10 | Treated as indicated § 63 disengages deep yellow vapors. ... *Bromate.* | 102 |
| | Violet vapors ... *Iodate.* | 102 |
| 11 | Treated as indicated § 65 imparts to the flame an azure-blue or green color | 12 |
| | Not | 17 |
| 12 | Heated in a matrass with bisulphate of potassa and a little peroxide of manganese disengages violet vapors. ... *Iodide and Iodate.* | 102 |
| | Deep yellow vapors, ... *Bromide and Bromate.* | 102 |
| | Not | 13 |
| 13 | Treated as indicated § 77 exhibits the fluorine-reaction. ... *Fluoride.* | 102 |
| | Not | 14 |
| 14 | Treated as indicated § 65 imparts to the flame an azure-blue color | 15 |
| | A green color | 16 |
| 15 | Heated with Sd on Ch gives a mass which, when mixed with bisulphate of potassa and black oxide of manganese and heated in a closed tube, evolves a deep yellow gas ... *Bromide.* | 102 |
| | Not ... *Chloride and Chlorate.* | 102 |
| 16 | Heated with Sd on Ch gives a mass which, when mixed with bisulphate of potassa and peroxide of manganese and heated in a closed tube, evolves violet vapors, *Iodide.* | 102 |
| | Deep yellow vapors ... *Bromide.* | 102 |
| 17 | Effervesces with hydrochloric acid ... *Carbonate.* | 102 |
| | Not | 18 |
| 18 | When finely powdered and heated with hydrochloric acid, effervesces ... *Carbonate.* | 102 |
| | Not | 19 |

| No. | Character | Go to |
|---|---|---|
| 19 | When finely powdered and heated with concentrated hydrochloric acid, gelatinizes..... | 30 |
| | Not | 20 |
| 20 | Fused with Sd on Ch yields neither a metallic globule nor a Ct | 21 |
| | Yields a metallic globule or a Ct | 22 |
| 21 | Treated as indicated § 61 colors the flame yellowish-green ..... *Borate.* | 102 |
| | Not | 23 |
| 22 | The scoriaceous mass is heated in a platinum spoon with a drop of concentrated sulphuric acid, then alcohol poured on it, and lighted. The flame appears yellowish-green, ..... *Borate.* | 102 |
| | Not, | 23 |
| 23 | Treated as indicated § 77 exhibits the fluorine-reaction, ..... *Fluoride.* | 102 |
| | Not | 24 |
| 24 | Heated with Sd on Ch yields a button of fused metal, | 25 |
| | Not | 27 |
| 25 | Heated on Ch alone behaves as indicated § 96; yields with Sd on Ch a soft globule ...... *Phosphate of Lead.* | |
| | Not | 26 |
| 26 | The scoriaceous mass treated with boracic acid, as indicated § 95, exhibits the reaction of phosphoric acid, *Phosphate.* | 102 |
| | Not | 28 |
| 27 | Treated as indicated § 95 exhibits the reaction of phosphoric acid ..... *Phosphate.* | 102 |
| | Not | 28 |
| 28 | With Sd on Ch yields a metallic button or a copious Ct | 31 |
| | Not | 29 |
| 29 | Pulverized and fused with 5 or 6 times its weight of Sd in a platinum spoon, yields a mass which, when heated with hydrochloric acid, gives a gelatinous precipitate | 30 |
| | Not | 31 |
| 30 | The gelatinous precipitate placed, while still moist, on a blade of iron or zinc, becomes blue ......... *Tungstate.* | 139 |
| | Not | 140 |
| 31 | Metallic aspect | 32 |
| | Not | 57 |

| No. | Characters | Go to |
|---|---|---|
| 32 | Yields on Ch a malleable and fusible metallic button, which is not oxidable | 33 |
| | Not | 36 |
| 33 | Yellow button | 34 |
| | White button | 35 |
| 34 | With Bx on platinum wire gives a bluish glass ... *Gold with Copper.* | |
| | Not ... *Gold.* | |
| 35 | With Bx on platinum wire gives a bluish glass ... *Silver with Copper.* | |
| | Not ... *Silver.* | |
| 36 | With Bx on platinum wire gives a glass which is blue in both flames ... *Cobalt.* | |
| | Not | 37 |
| 37 | Gives with Bx the reactions of oxide of copper | 38 |
| | Not | 42 |
| 38 | Red and malleable metallic button ... *Copper.* | |
| | Not | 39 |
| 39 | Deposits on Ch a Ct, yellow while hot, white when cold | 41 |
| | Not. A yellowish, brittle alloy ... *Copper and Tin.* | |
| 41 | Malleable, yellow or reddish alloy ... *Copper and Zinc.* | |
| | White, malleable alloy ... *Copper, Zinc, Nickel.* | |
| 42 | Very fusible metallic button | 43 |
| 43 | Deposits on Ch a Ct | 44 |
| | No Ct deposited on Ch; exhibits the reactions of tin ... *Tin.* | |
| 44 | White Ct, very volatile | 45 |
| | Not | 46 |
| 45 | Yields on Ch a brittle globule, which exhibits the antimony reactions ... *Antimony.* | |
| | Not ... *Tellurium.* | |
| 46 | Metallic aspect, or powder assuming metallic lustre under the polishing steel | 47 |
| | Not | 50 |
| 47 | Infusible and inoxidable ... *Platinum.* | |
| | Oxidable | 48 |
| 48 | With Bx in OFl an amethyst-colored glass, *Manganese.* | |
| | Not | 49 |

| | | |
|---|---|---|
| 49 | After having been oxidized, exhibits with fluxes the iron-reactions.......... *Iron.* | |
| | After having been oxidized, exhibits with fluxes the nickel-reactions.......... *Nickel.* | |
| 50 | Yields with Sd on Ch in RFl a tin globule, *Oxide of Tin.* | |
| | Not.......... | 51 |
| 51 | With Bx on platinum wire a green glass in both flames, .......... *Chromic Iron.* | |
| | Not.......... | 52 |
| 52 | Yields with Sd on platinum foil in OFl a bluish-green mass | 53 |
| | Not.......... | 54 |
| 53 | Gives with Bx or SPh on platinum wire an amethyst-colored bead.......... *Oxide of Manganese.* | |
| | Not. Brown powder, *Tungstate of Iron and Manganese.* | |
| 54 | With S Ph on platinum wire in RFl gives a glass which, on cooling, becomes brownish-red and, when touched with tin, violet red.......... *Titaniferous Iron.* | |
| | Not; exhibits the iron-reactions.......... | 55 |
| 55 | Heated in a closed glass tube, yields water; powder yellow.......... *Hydrate of Sesquioxide of Iron.* | |
| | Yields no water.......... | 56 |
| 56 | Magnetic; powder black.......... *Magnetic Iron.* | |
| | Not; powder red.......... *Peroxide of Iron.* | |
| 57 | Affords with Sd on Ch in RFl a fusible metallic button, | 58 |
| | Not.......... | 66 |
| 58 | The button is malleable and inoxidable.......... | 59 |
| | Oxidable button.......... | 60 |
| 59 | Yellow button.......... *Oxide of Gold.* | |
| | White button.......... *Oxide of Silver.* | |
| 60 | Button with Ct.......... | 61 |
| | Malleable button without Ct.......... | 75 |
| 61 | The Ct is white and very volatile, *Oxide of Antimony.* | |
| | Not.......... | 62 |
| 62 | The Ct is yellow, and the button soft.......... | 63 |
| | The button is brittle.......... | 65 |
| 63 | A very small quantity affords with Bx or SPh in OFl a green glass.......... *Chromate of Lead.* | |
| | Not.......... | 64 |

64 { The substance is yellow or reddish, *Protoxide of Lead.*
The substance is red...... ......................... *Minium.*
The substance is brown............... *Deutoxide of Lead.*

65 { Affords with Bx or SPh in OFl a green bead.............
........................... ..........*Chromate of Bismuth.*
Not....................................... *Oxide of Bismuth.*

66 { Treated on Ch in OFl deposits a Ct, or vaporizes completely........................................................ 67
Not.......................................................... 70

67 { The Ct is white, and very volatile...*Oxide of Antimony.*
Not.......................................................... 68

68 { The Ct is brown.........................*Oxide of Cadmium.*
Not...... ................................................ 69

69 { The substance is red or yellow and affords, when heated in a closed glass tube, metallic mercury.................
...........................................*Oxide of Mercury.*
The substance is white, becomes yellow on heating, and on cooling white again.........................*Oxide of Zinc.*

70 { Affords with Bx a bead which is blue in both flames......
.............................................*Oxide of Cobalt.*
Not.......................................................... 71

71 { The Bx bead is green in both flames......................... 72
Not.......................................................... 77

72 { Soluble in water.............................................. 73
Insoluble in water............................................ 74

73 { The substance is orange-red......*Bichromate of Potassa.*
The substance is yellow... *Chromate of Potassa or Soda.*

74 { The substance is of semi-metallic aspect or grayish-black
............................................*Chromic Iron.*
The substance is a green powder............................
............................... *Sesquioxide of Chromium.*

75 { The button is white............................*Oxide of Tin.*
The button is red............... ................................ 76

76 { The substance is red or brown...... *Suboxide of Copper.*
The substance is black................*Protoxide of Copper.*

77 { The bead is green in OFl, and becomes reddish-brown in RFl........................................................ 76
Not.......................................................... 78

| No. | Reaction | |
|---|---|---|
| 78 | The bead is amethyst-colored in OFl | 79 |
| | Not | 80 |
| 79 | Gives off water when heated in a glass tube ... *Hydrated Oxide of Manganese.* | |
| | Not ... *Oxide of Manganese.* | |
| 80 | Heated alone on Ch in OFl becomes magnetic | 55 |
| | Not | 81 |
| 81 | Exhibits with SPh on platinum wire the uranium reactions ... *Oxide of Uranium.* | |
| | Not | 82 |
| 82 | Soluble in water, exhibiting alkaline reaction | 83 |
| | Not | 91 |
| 83 | Very soluble | 84 |
| | But little soluble | 88 |
| 84 | Heated on platinum wire, fuses readily and vaporizes | 85 |
| | Not | 87 |
| 85 | Heated on platinum foil, stains it dark-yellow ... *Lithia.* | |
| | Not | 86 |
| 86 | Heated on platinum wire, colors the flame pale violet ... *Hydrate of Potassa.* | |
| | Reddish-yellow; the outer flame becomes enlarged ... *Hydrate of Soda.* | |
| 87 | Moistened with a drop of hydrochloric acid, and heated on platinum wire, colors the flame pale-green ... *Baryta.* | |
| | Purple ... *Strontia.* | |
| 88 | Moistened with a drop of hydrochloric acid, and heated on platinum wire, colors the flame purple ... *Strontia.* | |
| | Not | 89 |
| 89 | Heated with SoCo, assumes a flesh-color ... *Magnesia.* | |
| | Not | 90 |
| 90 | Heated alone, becomes very luminous ... *Lime.* | |
| | Not; colors the flame pale-green ... *Baryta.* | |
| 91 | Heated with SoCo, assumes a fine blue color ... *Alumina.* | |
| | Not | 92 |
| 92 | Heated with SoCo, assumes a flesh-color ... *Magnesia.* | |
| | Not | 93 |
| 93 | Heated with SoCo, assumes a green color ... *Oxide of Zinc.* | |
| | Not | 94 |

94 Affords with SPh in OFl a colorless glass, which in RFl becomes blue........ 95
Not........ 97

95 Heated in a closed glass tube, evolves ammonia and becomes blue or green........ *Tungstate of Ammonia.*
Not........ 96

96 On Ch alone, infusible........ *Tungstic Acid.*
Fusible........ *Tungstate of Potassa or Soda.*

97 Exhibits with SPh the reactions of........ *Molybdic Acid.*
Not........ 98

98 Exhibits with SPh the reactions of pure... *Titanic Acid.*
Not........ 99

99 Affords with SPh in RFl a reddish-yellow glass; the intensity of the color increases on cooling........ 100
Not........ 101

100 The glass, when heated on Ch with tin, becomes violet. ........ *Titanic Acid, containing Iron.*

101 With Sd on Ch in RFl, affords a metallic powder attractable by the magnet........ *Oxide of Nickel.*
Not: affords with SPh in OFl a glass which, while hot, is red, and colorless when cold..... *Oxide of Cerium.*

NITRATES, CHLORATES, BROMATES, IODATES, CARBONATES, PHOSPHATES, BORATES, CHLORIDES, BROMIDES, IODIDES, OXIDES, HYDRATES.

102 Affords with Sd on Ch in RFl a fusible metallic button. 103
Not........ 109

103 The button is malleable and inoxidable........ 104
The button is oxidable........ 105

104 The button is yellow........ *Salt of Gold.*
The button is white........ *Salt of Silver.*

105 The button is red and malleable........ *Salt of Copper.*
Not........ 106

106 The button is white and malleable and forms no Ct.... ........ *Salt of Tin.*
Not........ 107

107 Forms a white and very volatile Ct... *Salt of Antimony.*
The Ct is orange-yellow........ 108

108 { The button is malleable . ..................... *Salt of Lead.*
The button is brittle. ............................ *Salt of Bismuth.*

109 { Treated with Sd on Ch in RFl, deposits a Ct............ 110
Not............................................................ 112

110 { The Ct is white and very volatile.... *Salt of Antimony.*
Not............................................................ 111

111 { The Ct is reddish-brown................ *Salt of Cadmium.*
The Ct is yellow while hot, and white when cold.......
............................................... *Salt of Zinc.*

112 { On Ch alone, affords a gray and infusible powder which, under the polishing steel, assumes metallic lustre....
....................................... *Salt of Platinum.*
Not............................................................ 113

113 { Heated with Sd in a closed glass tube affords a sublimate of mercury......................... *Salt of Mercury.*
Not............................................................ 114

114 { Heated with Sd in a closed glass tube disengages ammonia.................................. *Salt of Ammonia.*
Not............................................................ 115

115 { Gives with Bx or SPh beads which are blue in both flames...................................... *Salt of Cobalt.*
Not............................................................

116 { The beads are green in both flames.... *Salt of Chromium.*
Not............................................................ 117

117 { The bead exhibits the reactions produced by oxide of copper....................................... *Salt of Copper.*
Not............................................................ 118

118 { Affords with Sd on Ch a metallic powder, which assumes lustre by friction, and is attracted by the magnet....... 119
Not............................................................ 120

119 { Gives with Bx in RFl a bottle-green glass. *Salt of Iron.*
Gives with Bx in RFl a grayish glass.... *Salt of Nickel.*

120 { Gives with Bx in OFl an amethyst-colored bead. ......
....................................... *Salt of Manganese.*
Not............................................................ 121

121 { Infusible mass, which assumes, with SoCo, previous to fusion, a fine blue color. ............. *Salt of Alumina.*
A flesh-color.............................. *Salt of Magnesia.*
Not............................................................ 122

| No. | Character | |
|---|---|---|
| 122 | The watery solution gives a precipitate on addition of some Sd........ | 123 |
| | Not........ | 124 |
| 123 | Heated on platinum wire, colors the flame pale-green........ *Salt of Baryta.* | |
| | Colors the flame purple........ *Salt of Strontia.* | |
| | Not; but becomes very luminous........ *Salt of Lime.* | |
| 124 | Heated on platinum wire, colors the flame violet ........ *Salt of Potassa.* | |
| | Colors the flame reddish-yellow........ *Salt of Soda.* | |

## Sulphur Compounds.

| No. | Character | |
|---|---|---|
| 125 | Metallic aspect; sulphides........ | 126 |
| | Not ........ | 127 |
| 126 | The substance is calcined, and the metal detected by proceeding as indicated above, beginning with No. ...... | 102 |
| 127 | Sulphates, hyposulphates, sulphites, hyposulphites, sulphides prepared artificially by precipitation, and a few native sulphides........ | 128 |
| 128 | Heated with hydrochloric acid disengages: | |
| | Sulphuretted hydrogen........ *Sulphide.* | 130 |
| | Sulphurous acid........ | 129 |
| | Nothing........ *Sulphate or Hyposulphate.* | 130 |
| 129 | Hydrochloric acid produces a white precipitate of sulphur........ *Hyposulphite.* | 130 |
| | Not........ *Sulphite.* | 130 |
| 130 | The metal is detected, beginning with No........ | 102 |

## Arsenic Compounds.

| No. | Character | |
|---|---|---|
| 131 | Metallic aspect........ | 132 |
| | Not........ | 135 |
| 132 | Readily and completely volatilized on Ch........ | 133 |
| | Not........ | 134 |
| 133 | Gives a white and very volatile Ct........ *Arsenide of Antimony.* | |
| | Not........ *Arsenic.* | |

134 The substance, which is an arsenide or sulpharsenide, is thoroughly calcined, and then the metal detected as indicated above, beginning with No.................. 102

135 Wholly volatilized on Ch, and exhibiting the reactions of sulphur .................................................. 136
Not wholly volatilized, or exhibiting no sulphur-reaction, 137

136 The substance is yellow.........................*Orpiment.*
The substance is red...............................*Realgar.*

137 The substance is very volatile............*Arsenous Acid.*
Not: *arsenite* or *arsenate.* The substance is well calcined with alternating OFl and RFl, and the metal found, beginning with No.................................. 102

## Selenium Compounds.

138 Metallic aspect......................................*Selenide.*
Not: *selenite* or *selenate;* the substance is well calcined, and the metal detected, beginning with No.... 102

## Tungstates.

139 With Sd on platinum wire in OFl affords a greenish-blue mass ........................................*Wolfram.*
Not.................................................................. 140

140 Heated with Sd in a closed glass tube, evolves ammonia..............................*Tungstate of Ammonia.*
Not.................................................................. 141

141 Soluble.........................*Tungstate of Potassa or Soda.*
Insoluble.. ...............*Tungstate of Lime, Baryta, &c.*

## Silicates.

142 The analogy in chemical composition and properties, and the number of native silicates, make it impossible to discriminate them by a few simple tests.* The base or bases may, however, in many cases be detected by proceeding as indicated above, beginning with No.................................................. 102

* For the discrimination of the native silicates, v. Chapter VI.

# SIXTH CHAPTER.

## ON THE DISCRIMINATION OF MINERALS BY MEANS OF THE BLOW-PIPE, AIDED BY HUMID ANALYSIS.

By the methods given in the preceding chapters, we can readily detect the constituents of most inorganic compounds, whether prepared artificially or occurring in nature; especially if heavy metals form the principal constituents. But these methods do not enable us to discriminate the different native silicates, and other mineral bodies, which consist essentially of such substances that do not show any very characteristic reactions before the blow-pipe, as ex. gr. the alkaline earths. In some cases we may succeed in ascertaining the principal ingredients of the substance under examination, but fail in establishing the mineral species. To attain this end more securely, we must pursue a course, composed of an examination of the physical properties of the body and of blow-pipe operations, aided by humid analysis. The course adopted in this "Manual" is that given by *Franz von Kobell,* as laid down in his "*Tafeln zur Bestimmung der Mineralien.*" The following is only an extract, slightly modified, from this treatise:

The minerals, according to Von Kobell's system, are arranged in two large groups, the first embracing those possessing metallic lustre, the second those devoid of metallic lustre. To avoid mistakes, originating in the fact that some minerals occur sometimes with, and sometimes without, metallic lustre, these minerals will be found enumerated in both groups.

The same precaution has been taken in regard to those species in which degree of fusibility, whether below or above 5, might appear doubtful. The degree of fusibility is to be judged of from the following scale:

1. Gray Antimony.—Fusible in coarse splinters in the flame of a candle.
2. Natrolite.—Fusible in fine splinters in the flame of a candle.

3. Almandine or Iron-Garnet.—Easily fusible before the blowpipe.

4. Actinolite (a variety of hornblende).—Fusible before the Blp in coarse splinters.

5. Orthoclase.—Fusible before the Blp in fine splinters.

6. Broncite.—Fusible on the edges in very fine splinters.

The fusibility, when equal to that of actinolite, is designated by 4 ; when between that of natrolite and almandine, by 2, 5, and so on.

The two large groups are divided into classes according to the fusibility ; these again in divisions, &c., by which means we obtain the following general classification:

## Group I. Minerals possessing a Metallic Lustre.

Class I. Native malleable metals, and mercury.

Class II. Fusibility 1—5, or readily volatile.

- Division 1. Give a strong arsenical odor on Ch.
- Division 2. Give on Ch, or in an open tube, the horse-radish odor of selenium.
- Division 3. Give in an open tube a white or grayish sublimate, which is fusible into colorless drops, indicative of tellurium.
- Division 4. Give antimonial fumes on Ch.
- Division 5. Give with Sd on Ch a sulphur-reaction, but do not give indications as above.
- Division 6. Do not exhibit the properties of the preceding divisions.

Class III. Infusible, or fusibility above 5, and not volatile.

- Division 1. Give with Bx, in small quantities, the manganese-reaction.
- Division 2. Treated on Ch in RFl, become magnetic.
- Division 3. Resembling those of division 2.

## Group II. Minerals not possessing Metallic Lustre.

Class I. Easily volatile, or combustible.

Class II. Fusibility 1—5, not, or only partially, volatile.

- Part I. Give with Sd on Ch a metallic globule or magnetic metallic mass.

Division 1. Give with Sd a globule of silver.

Division 2. Give with Sd a globule of lead.

Division 3. When moistened with hydrochloric acid, color the flame blue, and give with nitric acid a solution which, on addition of an excess of ammonia, assumes an azure-blue color.

Section 1. Give on Ch a strong arsenical odor.

Section 2. Give no arsenical odor.

Division 4. Impart to the Bx bead a blue color.

Division 5. When fused on Ch in RFl, give a black or gray metallic magnetic mass.

Section 1. Give on fusion a strong arsenical odor.

Section 2. Soluble in hydrochloric acid without leaving a perceptible residue, and without gelatinizing.

Section 3. With hydrochloric acid, form a jelly, or are decomposed with separation of silica.

Section 4. But little affected by acids.

Division 6. Not belonging to either of the preceding divisions.

Part II. With Sd on Ch, give no metallic globule, or magnetic metallic mass.

Division 1. After fusion and continued heating on Ch or in the forceps, have an alkaline reaction, and change to blue the color of a moistened red litmus-paper.

Section 1. Easily and completely soluble in water.

Section 2. Insoluble in water, or soluble with difficulty.

Division 2. Soluble in hydrochloric acid without leaving a perceptible residue, some also soluble in water; not gelatinizing.

Division 3. Soluble in hydrochloric acid, forming a perfect jelly.

Section 1. Giving water in a matrass.

Section 2. Giving traces, or no water in a matrass.

Division 4. Soluble in hydrochloric acid with separation of silica, without forming a perfect jelly.

Section 1. Giving water in a matrass.

Section 2. Giving traces, or no water in a matrass.

Division 5. Little affected by hydrochloric acid; imparting to the Bx bead the color of manganese.

Division 6. Not belonging to either of the preceding divisions.

CLASS III. Infusible, or fusibility above 5.

Division 1. After ignition moistened with SoCo and again ignited, assume a bright-blue color.

Section 1. Giving much water in a matrass.

Section 2. Giving little or no water in a matrass.

Division 2. Moistened with SoCo and ignited, assume a green color.

Division 3, After ignition have an alkaline reaction, and turn into blue the color of a moistened red litmus-paper.

Division 4. Completely soluble, or nearly so, in hydrochloric or nitric acid, without gelatinizing or leaving a perceptible residue of silica.

Division 5. With hydrochloric acid, form a jelly or are decomposed with separation of silica.

Section 1. Giving water in a matrass.

Section 2. Giving traces, or no water in a matrass.

Division 6. Not belonging to either of the preceding divisions.

Section 1. Hardness below 7.

Section 2. Hardness=7, or above.

## GROUP I. MINERALS POSSESSING A METALLIC LUSTRE.

CLASS I. NATIVE MALLEABLE METALS, AND MERCURY.

Native silver, see § 197.

Native Gold and Electrum (alloy of silver and gold), see § 150.

Native copper, see § 134.

Native Lead, characterized by coating on charcoal (see § 23) and softness; H=1.5.

Native Platinum, see § 152.

Native Palladium, distinguished from the preceding by being soluble in nitric acid.

Native Iron, see § 154.

Native Mercury, see § 187.

CLASS II. FUSIBILITY 1 TO 5, OR READILY VOLATILE.

*Division 1. Give a strong arsenical odor on charcoal.*

Native Arsenic, see § 118.

Dufrenoysite, see § 170; Tennantite, see § 139; Polybasite, see § 207; Domeykite, see § 140.

Smaltine, see § 128; Cobaltine, see § 129.

Copper Nickel, see § 191; Gersdorffite, see § 192; Chloanthite=NiAs, resembles the preceding two; distinguished from copper nickel by its tin-white color, and from gersdorffite by not giving the reactions for sulphur.

Arsenical Pyrites, see § 161.

*Division* 2. *Give on charcoal, or in an open tube, the horse-radish odor of selenium.*

Selenide of mercury=$HgSe^2$, Onofrite=Hg (S.Se) and selenide of mercury and lead (Selenquecksilberblei)=3PbSe+HgSe, yield metallic mercury on being heated with Sd in a closed glass tube (§ 91); the latter yields a globule of metallic lead on being heated on charcoal with Sd.

Clausthalite=PbSe. Color lead-gray; volatile without previous fusion, depositing first a slight gray, then a white, and finally a greenish-yellow coating; with Sd yields with difficulty globules of lead.

Naumannite=AgSe. Color iron-black; melts readily and yields with Bx a globule of pure silver.

Berzelianite=$Cu^2Se$ and Eucairite $Cu^2Se$+AgSe. Color of the former silver-white, of the latter lead-gray. Distinguished from the other minerals of this division by giving copper-reactions.

*Division* 3. *Give in an open tube a white or grayish sublimate, which is fusible into colorless drops, indicative of tellurium, see* § 11.

The assay-piece used for this experiment ought not to be very small. It must also be borne in mind that the minerals of this division frequently evolve an odor of selenium, owing to a small percentage of selenium which they contain as adventitious constituent.

The minerals of this division may be subdivided according to their color.

*a.* Ores of tellurium of tin-white or silver-white color.

Native Tellurium, fuses readily and is volatile without leaving a residue.

Hessite=AgTe, and Altaite=PbTe; both soluble in nitric acid; the former yields with Sd on Ch a globule of metallic silver.

Some varieties of sylvanite, see § 151.

*b.* Ores of tellurium of lead-gray or steel-gray color.

Tetradymite, see § 123.

Sylvanite, see § 151.

Nagyagite=Pb, Au, Te, S. Color blackish lead-gray. Distinguished from the preceding by its solution in nitric acid giving a copious precipitate with sulphuric acid.

*Division* 4. *Give copious antimonial fumes on charcoal* (*see* § 16).

The fumes possess sometimes the odor of sulphurous acid or arsenic.

Native Antimony, distinguished by its tin-white color; Stibnite, see § 115; Zinkenite, see § 170; Jamesonite, see § 170; Bournonite, see § 170.

The powdered stibnite, on being heated with hydrate of potassa, assumes a yellow color, while the latter three minerals, which are steel-gray, do not change color. Bournonite, on being treated with nitric acid, imparts to the solution a sky-blue color, and gives copper-reactions on being treated as described in § 71. Zinkenite and jamesonite are converted into white powders by treatment with nitric acid without imparting a color to the acid; they are distinguished by their hardness, that of zinkenite being=3.5, that of jamesonite=2.5.

Closely resembling the above in their chemical behavior are the following rare minerals: Plumosite, see § 170; Boulangerite, see § 120; Geokronite, see § 170; Plagionite, see § 170; Kilbrikenite; Steinmannite.

Discrasite, see § 198; Stephanite, see § 206; Polytelite=S, Sb, Zn, Fe, Ag, Cu; some varieties of Tetrahedrite, see § 138; Miargyrite=AgS, $SbS^3$. Discrasite does not give a sulphur-reaction, all the others do. Polytelite gives a copper-reaction on being treated as described in § 73. Miargyrite, streak dark cherry-red; stephanite, streak black. Miargyrite and stephanite, hardness=2.5; Polytelite, hardness=3.5. All the minerals of this subdivision give a globule of silver on being treated as described § 104 or § 105.

Wolfsbergite (antimonial copper)=$Cu^2S$, $SbS^3$; does not give a globule of silver, but yields a globule of metallic copper on being treated with Sd on charcoal.

Ullmannite, see § 193; Berthierite, see § 116; Breithauptite=Ni Sb. All yield a magnetic globule with continued heat. Breithauptite is distinguished from the other two by not giving a sulphur-reaction.

*Division* 3. *Give with Sd on Ch a sulphur-reaction, but do not give the general reactions of the preceding divisions.*

Silver Glance, § 203.

Galena, see § 169.

Cinnabar, see § 90.

Manganblende=MnS. Color iron-black, streak green. The pulverized mineral evolves sulphuretted hydrogen with hydrochloric acid.

Hauerite=$MnS^2$. Color brownish-black, streak brownish-red. Yields sulphur on being heated in a matrass.

Copper Glance, see § 137; Stromeyerite, see § 208; Tin Pyrites, see § 210; Copper Pyrites, see § 135; Purple Copper, see § 136; Cuban $=Cu^2S, Fe^2S^3$; Wittichite $=3Cu^2S, BiS^3$; Aikinite (acicular bismuth) $=3Cu^2S, BiS^3+2(3PbS,BiS^3)$; Grunauite $=BiS^3+10Ni^2S^3$; Cuproplumbite $=Cu^2S, 2PbS$. All these minerals are partially soluble in nitric acid, the solution possessing a sky-blue or green color; on addition of water to the concentrated solution a white precipitate is produced, if the mineral under examination was wittichite, grunauite, or aikinite. [To distinguish these three, add to the acid solution sulphuric acid: a precipitate indicates aikinite; wittichite gives the copper-reaction on being treated as described in § 73, grunauite not.] Copper pyrites and cuban are distinguished from the others by their brass-yellow color; purple copper is also characterized by its color. To distinguish the remaining four minerals, make a solution in nitric acid; add sulphuric acid: a precipitate indicates cuproplumbite; if no precipitate is produced, add hydrochloric acid: a precipitate indicates stromeyerite; to distinguish between copper-glance and tin pyrites, see § 137 and § 210.

Millerite, see § 194; Linnæite, see § 130; Iron Pyrites, see § 158; Marcasite, see § 159; Sternbergite $=$ S, Ag, Fe. The members of this subdivision fuse to globules which are attracted by the magnet. They are readily distinguished by the characteristics given in Chapter III. Sternbergite, by the treatment described in § 104, yields a globule of silver. Marcasite and iron pyrites can only be distinguished by their crystalline form.

Bismuthine, see § 125.

*Division* 6. *Do not exhibit the properties of the preceding divisions.*

Amalgam, see § 188.

Native Bismuth, see § 122.

Hematite, see § 156.

Magnetite, see § 157.

Wolfram=$MnO,FeO,Wo^3$. Color dark grayish or brownish-black. Fusibility=3. The pulverized mineral on being boiled with aqua regia assumes gradually a yellowish color.

Samarskite=$NbO^3$, $FeO$, $U^2O^3$, $YO$. Color velvet-black. Fusibility=4. By fusing the pulverized mineral with hydrate of potassa, boiling the fused mass with hydrochloric acid, filtering, and concentrating the solution by boiling with addition of tin foil: the liquid assumes a fine blue color which does not pass into red on addition of water (as is the case with compounds containing titanium), but becomes paler and gradually disappears.

Rhodonite, dark varieties=$3MnO.SiO^3$, $3HO$. Yields water on being heated in a matrass. Soluble in hydrochloric acid with separation of silica.

Some varieties of psilomelane, see § 183.

Lievrite and Allanite, some varieties, see p. 123.

Plattnerite=$PbO^2$. Color iron-black; easily reduced to metallic lead.

Red Copper, some varieties, see § 142.

CLASS III. INFUSIBLE, OR FUSIBILITY ABOVE 5.

*Division* 1. *Give with borax, in small quantities, the manganese reactions.*

The members of this division are distinguished from each other principally by their physical properties.

Braunite, see § 182; Hausmannite, see § 181; Psilomelane, see § 183; Pyrolusite, see § 180; Franklinite, some varieties, see § 186; Manganite=$Mn^2O^3$, $HO$. Color steel-gray to iron-black; streak dark reddish-brown; hardness 3—4; yields water in a matrass.

*Division* 2. *Heated on charcoal in reduction-flame, become magnetic.*

Hematite, see § 156.

Franklinite, see § 186; Magnetite, see § 157.

Titaniferous iron, see § 162; some varieties of Rutil and Arkansite (see below); some varieties of Limonite (§ 155), and Blende (§ 212).

*Division* 3. *Minerals resembling those of Division* 2.

Chromic Iron, see § 127.

Molybdenite=$MoS^2$; Graphite=C. Both very soft, hardness=1.5. Molybdenite, when heated in the forceps, colors the flame greenish; and gives a sulphur-reaction when treated as described in § 107.

Arkansite=$TiO^2$; Perofskite=$CaO.TiO^2$. Both give the reaction for titanium as described § 111. Distinguished by crystalline form.

Iridosmine, see § 153.

Tantalite and Columbite=MnO, FeO, $TaO^3$, $NbO^3$, $WO^3$, $SnO^2$; Yttro-tantalite=3(CaO.YO.FeO), ($TaO^3$, $WO^3$). The color of these minerals is iron-black; yttro-tantalite loses its color before the Blp and becomes yellowish or white, that of the others remains unchanged. Acids affect them but little. Tantalite and columbite give the same reaction as samarskite when treated with hydrate of potassa, &c. (See p. 117.)

Pitchblende. Color usually velvet-black, lustre greasy; partially soluble in nitric acid to a yellow liquid; the solution gives a sulphur-yellow precipitate with ammonia.

## GROUP II. MINERALS NOT POSSESSING METALLIC LUSTRE.

### CLASS I. EASILY VOLATILE OR COMBUSTIBLE.

Native Sulphur. Completely volatile, burns with a blue flame and emission of sulphurous acid.

Realgar, see § 119; Orpiment, see § 120.

Arsenolite, see § 121.

Red Antimony, see § 117; Valentinite=$SbO^3$. Color white; does not change color with hydrate of potassa; does not evolve sulphuretted hydrogen with hydrochloric acid.

Sal-ammoniac=$NH^4Cl$; Mascagnine=$NH^3,SO^3+2HO$. Color white; both evolve ammonia with hydrate of potassa; the former is volatile without previous fusion, the latter intumesces.

Cinnabar, see § 190; Calomel, see § 189.

CLASS II. FUSIBILITY 1—5; NOT, OR ONLY PARTIALLY, VOLATILE.

Part I. Give with carbonate of soda on charcoal a metallic globule or a magnetic metallic mass.

*Division* 1. *Give with carbonate of soda a globule of silver.*

Proustite, see § 205; Ruby Silver, see § 204; Xanthocone=$3AgS, AsS^5+2(3AgS, AsS^3)$, behaves like proustite, from which it is distinguished by its yellow streak.

Horn Silver, see § 199; Iodyrite, see § 202.

Selbite=$AgO, CO^2$, dissolves in nitric acid with effervescence.

*Division* 2. *Give with carbonate of soda a globule of lead.*

The minerals of this division are all soluble in nitric acid; the solution gives a copious precipitate with sulphuric acid.

Mimetine= $PbCl+3(3PbO, AsO^5)$; Hedyphane= $PbCl+3(3[PbO.CaO], [AsO^5.PO^5])$. The former completely, the latter partially reduced to metallic lead with evolution of arsenical fumes.

Pyromorphite, see § 175.

Minium, see § 168; Crocoisite, see § 177; Melanochroite=$3PbO,2CrO^3$. Aræoxene=$VO^3, AsO^5, PbO, ZnO$. Crocoisite and melanochroite give the chromium-reaction (§ 67). The latter three, on being boiled with hydrochloric acid, give an emerald-green solution; on adding alcohol to the liquid, concentrating by heat, pouring off from the residue, and then adding water: the liquid assumes a sky-blue color if the mineral was aræoxene.

Linarite=$PbO.SO^3+CuO.HO$ is characterized by its deep azure-blue color.

Cerusite, see § 172; Cerasine, see § 171; Leadhillite, see § 173; Lanarkite = $PbO.CO^2+PbO.SO^3$. All soluble in nitric acid with effervescence; leadhillite and lanarkite leave an insoluble residue of sulphate of lead. The solution of cerasine gives with nitrate of silver a precipitate of chloride of silver.

Mendipite=$PbCl+2PbO$; Matlockite=$PbCl+PbO$. Dissolve in nitric acid without effervescence; the solution gives a precipitate with nitrate of silver.

Anglesite, see § 174.

Wulfenite, see § 129.

Scheeletine=$PbO,WO^3$. Color yellow, yellowish-brown, lustre resinous. Soluble in abundant quantity of hydrochloric acid, leaving a yellowish-green residue ($WO^3$). With sulphuric acid the pulverized mineral assumes a bright lemon-yellow color.

Vauquelinite, see § 178; Vanadinite=$2\ PbO, VO^3$ with $PbCl+2PbO$. Color of the former blackish-green, olive-green; of the latter brown, yellowish. Both impart to the borax-bead an emerald-green color. Both are soluble in nitric acid; the solution of vanadinite is yellow and gives a precipitate with nitrate of silver. That of vauquelinite not.

*Division* 3. *When moistened with hydrochloric acid, color the flame blue; and give with nitric acid a solution which, on addition of an excess of ammonia, assumes an azure-blue color.*

Section 1. Give on charcoal a strong arsenical odor.

Olivenite, see § 147.

Tyrolite, see § 148; Chalcophyllite=$8CuO.AsO^5$+ 23HO. Color green. Both decrepitate violently and yield much water; chalcophyllite dissolves in ammonia without leaving a residue.

Liroconite=$AsO^5$,$PO^5$,CuO,$AlO^3$,HO. Color sky-blue. Does not decrepitate; loses 22 per cent. of water on ignition.

Euchroite = $4CuO.AsO^5$ + 7HO; Erinite = 5CuO. $AsO^5$+2HO. Color of both emerald-green. The former loses by ignition 18½ per cent. of water, the latter only 5 per cent.

Section 2. Do not give an arsenical odor on charcoal.

Atakamite, see § 141.

Cyanosite, see § 145; Brochantite = $3CuO.SO^3$ + 3HO; Covelline=CuS. These three minerals give a sulphur-reaction (§ 107); cyanosite is soluble in water, the other two not. Color of covelline dark indigo-blue, of brochantite emerald-green.

Red Copper, see § 142; Melaconite=CuO; Tenorite=CuO. The color of the latter two is dark steel-gray to black. All three dissolve readily in acids without effervescence (except impure varieties of melaconite).

Malachite, see § 143; Azurite, see § 144; Mysorin =$CuO.CO^2$. Color blackish-brown; does not yield water in a matrass. All three dissolve readily in acids with effervescence.

Phosphocalcite, see § 146; Libethenite=$4CuO.PO^5$ +HO; Ehlite=$5CuO.PO^5$+3HO; Tagilite=4 $CuO.PO^5$+3HO. Are all readily soluble in nitric acid without effervescence; the (slightly acid) solution gives a precipitate with acetate of lead. Phosphocalcite loses 14 per cent. of water on ignition, the others less (from 7—10½).

11

Libethenite is dark olive-green; ehlite and tagilite emerald-green.

Chalcolite=$3CuO, PO^5 + 2(U^2O^3, PO^5) + 24 HO$. Color emerald-green. Dissolves in nitric acid to a yellowish-green liquid; on addition of ammonia in excess, a bluish-green precipitate is formed, the supernatant liquid being blue.

*Division* 4. *Impart to the borax-bead a blue color.*

Erythrine, see § 131; Annabergite, see § 196.

*Division* 5. *When fused on charcoal in reduction-flame, give a black metallic magnetic mass.*

To observe well the magnetic character of the fused mineral, it is advisable to expose a pretty large assay-piece to the action of the reduction-flame.

Section 1. Evolve a strong arsenical odor on being fused.

Scorodite, see § 166; Pitticite=$Fe^2O^3, AsO^5+HO$; Beudantite=$3 FeO.AsO^5+3 Fe^2O^3, 2 AsO^5+18 HO$. The pulverized minerals assume with hydrate of potassa a reddish-brown color. Scorodite and beudantite occur crystallized; pitticite massive, reniform.

Arseniosiderite=$5CaO,AsO^3+3(2FeO,AsO^5)+11 HO$. Color yellowish-brown; fibrous; lustre silky.

Pyromeline=$NiO, SO^3, HO, AsO^5$. Partly soluble in water; the solution assumes a blue color on addition of ammonia.

Section 2. Soluble in hydrochloric acid without leaving a perceptible residue, and without gelatinizing.

Green Vitriol, see § 164; Botryogen=$3FeO, 2SO^3 +3(Fe^2O^3,2SO^3)+36HO$. Are soluble in water; botryogen leaves an ochreous residue. A similar behavior show Copiapite=$2Fe^2O^3, 5SO^3+18HO$ (color yellow), and Coquimbite=$Fe^2O^3, 3SO^3+ 9HO$, color white.

Spathic Iron, see § 163.

Hureaulite=$3(5MnO,2PO^5)+5FeO,2PO^5+30HO$; Triplite=$4MnO.PO^5+4FeO.PO^5$. Fuse readily;

moistened with sulphuric acid give the phosphoric acid reaction (§ 35); with borax strong manganese-reaction; hureaulite yields much water, triplite none or very little.

Triphiline=3 LiO, $PO^5$+6(3[FeO, MnO], $PO^5$) shows a similar behavior; the manganese-reaction is less decided. On dissolving the mineral in hydrochloric acid, evaporating the solution to dryness, adding alcohol, heating the alcohol to ebullition and burning the vapor, the flame assumes a purple color.

Vivianite, see § 165; Anglarite=4 FeO.$PO^5$+4HO; Dufrenite=2 $Fe^2O^3$.$PO^5$+2½ HO; Cacoxene=2 FeO.$PO^5$+12HO. Fuse readily and behave with sulphuric acid like the preceding; give no manganese-reaction. Yield much water in a matrass: cacoxene 33 per cent.; vivianite 28 per cent.; anglarite 16 per cent.; dufrenite 8½ per cent. Color of anglarite bluish-gray; of dufrenite leek-green; of cacoxene ochre-yellow.

Hematite, see § 156.

Section 3. With hydrochloric acid form a jelly, or are readily decomposed with separation of silica.

Cronstedtite = 3(FeO.MnO.MgO), $SiO^3$ + $Fe^2O^3$, 3HO. Color black; streak dark leek-green; yields water; gelatinizes with hydrochloric acid.

Lievrite=3(3[FeO, CaO], $SiO^3$)+2($Al^2O^3$, $Fe^2O^3$), $SiO^3$; Allanite=3(CeO, CaO), $SiO^3$+2([$Fe^2O^3$, $Al^2O^3$], $SiO^3$). Yield no water, or only a trace; gelatinize with hydrochloric acid; allanite fuses with intumescence to a voluminous brownish or blackish glass; lievrite intumesces but slightly, decrepitates and fuses to an iron-black bead. Hardness of allanite=6, of lievrite=5—6.

Pyrosmalite=$Fe^2Cl^3$+$Fe^2O^3$,6HO+4([3FeO,2$SiO^3$]+[3 MnO, 2 $SiO^3$]). Does not gelatinize; fusibility=2; gives the chlorine-reaction (§ 65).

Allochroite [lime-iron-garnet] $= 3CaO,SiO^3 + Fe^2O^3, SiO^3$. Gelatinizes imperfectly; fuses readily; distinguished from the preceding by absence of cleavage.

Hisingerite $= (3FeO,Fe^2O^3), SiO^3 + xHO$; Xylotile [a variety of serpentine]. Fuse with difficulty; do not gelatinize. The former is black, amorphous; the latter brown, fibrous, woody. Both yield water in a matrass.

Some impure varieties of Limonite, see § 155.

Section 4. But little affected by acids.

Crocidolite $= 3(NaO,MgO), 4SiO^3 + 3(3FeO, 2SiO^3) + xHO$; Arfvedsonite $= NaO,SiO^3 + 3FeO,2SiO^3$. Fusibility $= 1.7$—2. Color of crocidolite lavender-blue or leek-green, fibrous, yields water in a matrass; arfvedsonite is black and yields no water.

[See also Hornblende and Tourmaline, below, some varieties of which become slightly magnetic after fusion.]

Green Earth [a variety of pyroxene]. Fusibility $= 3$; color celandine-green; hardness $= 1$; earthy.

Acmite $= NaO.SiO^3 + Fe^2O^3,2SiO^3$; Hedenbergite [a black pyroxene] $= 3CaO.2SiO^3 + 3FeO.2SiO^3$. Fusibility of the former $= 2$, of the latter $= 2.6$; form a black lustrous slag. Both are cleavable.

Almandine [iron-garnet] $= 3FeO.SiO^3 + Al^2O^3.SiO^3$. Fusibility $= 3$; hardness $= 7$—7.5. Color red, reddish-brown. Not cleavable.

Rhodonite, some varieties; see below.

Lepidolite, some varieties; see below.

*Division* 6. *Not belonging to either of the preceding divisions.*

Molybdine $= MoO^3$. Color sulphur-yellow; earthy. Gives with the fluxes the reactions of molybdic acid. Dissolves readily in hydrochloric acid; the solution is colorless, but turns blue on being stirred with an iron spatula.

Eulytine $= 2\ BiO^3, 3\ SiO^3$ with some phosphate and

fluoride of iron. Gelatinizes with hydrochloric acid. On charcoal yields a globule of metallic bismuth.

Bismutite, see § 124.

Part II. With carbonate of soda on charcoal, give no metallic globule or magnetic metallic mass.

*Division* 1. *After fusion and continued heating on charcoal or in the forceps, have an alkaline reaction, and change to blue the color of a moistened red litmus-paper.*

Section 1. Readily and completely soluble in water.

Nitre=$KO.NO^5$; Nitratine=$NaO.NO^5$. Deflagrate vividly on burning coals. Fused on platina wire, the former colors the flames bluish with a red tint; the latter bright-yellow.

Natron=$NaO.CO^2+10HO$; Trona=$2NaO.3CO^2+4HO$. The watery solution has an alkaline reaction, and effervesces on addition of hydrochloric acid.

Glauber Salt=$NaO.SO^3+10HO$; Thenardite=$NaO.SO^3$; Glaserite=$KO.SO^3$; Epsomite=$MgO.SO^3+7HO$; Potash Alum=$KO.SO^3+Al^2O^2.3SO^3+24HO$. The watery solutions of these minerals give a copious precipitate with chloride of barium; the solution of potash alum and epsomite are precipitated by carbonate of potassa [distinguished by reaction with solution of cobalt, § 44]; the concentrated solution of glaserite gives a precipitate with bichloride of platina; glauber salt yields much water, thenardite none.

Common Salt=$NaCl$. The watery solution gives a copious precipitate with nitrate of silver. Gives also the reactions for chlorine described §§ 65, 66.

Borax=$NaO.2BO^3+10HO$. Gives the reaction for boracic acid, § 60.

Section 2. Insoluble in water, or soluble with difficulty.

Gay-Lussite$=CaO.CO^2+NaO.CO^2+6HO$; Whitherite$=BaO.CO^2$. Dissolve in dilute hydrochloric acid with effervescence; the former yields water, the latter not.

Anhydrite$=CaO.SO^3$; Gypsum$=CaO.SO^3+2HO$; Polyhalite $= KO.SO^3 + MgO.SO^3 + 2(CaO.SO^3) + 2HO$; Glauberite $= NaO.SO^3 + CaO.SO^3$. Soluble in much hydrochloric acid; in the solution chloride of barium gives a precipitate. Gypsum yields much water, polyhalite little, the rest none; anhydrite is distinguished by superior hardness$=3.5$; polyhalite is distinguished from glauberite by its solution giving a yellow precipitate with bichloride of platina.

Barytes$=BaO.SO^3$; Celestine$=SrO.SO^3$. Insoluble in hydrochloric acid; give a sulphur reaction when treated as described § 107. Celestine colors the flame red, § 34; barytes yellowish-green, § 35.

Fluor$=CaF$; Cryolite$=3NaF+Al^2F^3$; Pharmacolite$=2CaO.AsO^5+6HO$. Do not effervesce with acids, and give no sulphur reaction. Pharmacolite evolves arsenical odor on charcoal; the other two give fluorine reaction, § 76. Fusibility of fluor$=3$, of cryolite$=1$.

Chiolite$=3NaF+2Al^2F^3$, behaves like cryolite; occurs only massive granular; white cryolite is distinctly crystalline and cleavable in 3 directions.

*Division* 2. *Soluble in hydrochloric acid without leaving a perceptible residue; some also soluble in water; not gelatinizing.*

Ammonia Alum $= NH^4O. SO^3 + Al^2O^3. 3SO^3 + 24HO$; Goslarite$=ZnO.SO^3+7HO$. Both soluble in water; give sulphur reaction, § 107. Heated on charcoal and treated with solution of cobalt, the former assumes a blue, the latter a green, color, §§ 44, 45.

Sassolin=$BO^3,3HO$ ; Boracite=$3MgO,4BO^3$ ; Hydroboracite=$3(CaO.MgO),4BO^3+18HO$. Give the boracic acid reaction, § 60. Sassolin is soluble in alcohol, the others not; boracite yields no water, while the others do.

Manganblende and Hauerite give strong manganese reaction; see p. 115.

Wagnerite=$MgF+3MgO.PO^5$; Apatite=$3(3CaO.PO^5)+Ca(Cl,F)$. Moistened with sulphuric acid, impart a pale bluish-green color to the flame. Fusibility of wagnerite=3—3.5 (with intumescence); of apatite=5 (without intumescence); wagnerite is soluble in dilute sulphuric acid; apatite not.

Amblygonite = $LiO$, $NaO$, $Al^2O^3$, $PO^5$, F. Fusibility=2; hardness=6. With difficulty soluble in concentrated sulphuric or hydrochloric acid.

Uranite = $3CaO.PO^5 + 2(3U^2O^3.PO^5) + 24HO$. Fuses readily, yields water, and gives with fluxes the reactions of sesquioxide of uranium. See Table II.

*Division* 3. *Soluble in hydrochloric acid, forming a perfect jelly.*

Section 1. Give water in a matrass.

Datholite = $3(CaO.BO^3) + 3CaO.4SiO^3 + 3HO$. Yields but little water, and gives the boracic acid reaction, § 60.

Natrolite=$NaO.SiO^3+Al^2O^3.SiO^3+2HO$. Fusibility=2, does not intumesce; hardness=5—5.5.

Scolecite=$CaO.SiO^3+Al^2O^3.SiO^3+3HO$; Laumontite = $3CaO. 2SiO^3 + 3(Al^2O^3. 2SiO^3) + 12HO$. Scolecite, on being heated, curls up like a worm and finally melts to a bulky, shining slag, which in the inner flame becomes a vesicular slightly translucent bead; hardness=5.5. Laumontite intumesces and fuses to a white translucent enamel; hardness=3.

Nearly related to scolecite and showing a similar behavior, are Mesolite and Thomsonite.

Phillipsite = $(CaO,KO).SiO^3 + Al^2O^3.SiO^3 + 5HO$. Fusibility=3, with slight intumescence; occurs usually in twin crystals.

Section 2. Giving only traces or no water in a matrass.

Helvin=$MnO, MnS + 2 (MnO. BeO. FeO), SiO^3$; Tephroite=$3MnO.SiO^3$. Distinguished from the other minerals of this section by giving manganese-reactions. Color of helvin wax-yellow, hardness=6—6.5; of tephroite ash-gray, hardness=5.5—6.

Hauyne and Lapis Lasuli=$SiO^3,Al^2O^3$, CaO, KO, $SO^3$,S, are of azure-blue color; give sulphur-reaction, § 107. Fusibility of the former=4.5, of the latter=3.

Nosean and Skolopsite=$SiO^3$, $Al^2O^3$, CaO, NaO, $SO^3$, of gray or brownish color; give sulphur-reaction, § 107. Fusibility of nosean=4.5; of skolopsite=3 (with intumescence like idocrase).

Sodalite = $NaCl + 3 NaO. SiO^3+3 (Al^2O^3, SiO^3)$; Eudialyte=$SiO^3$, $ZrO^3$, CaO, NaO, FeO, Cl, give the chlorine-reaction, § 65. The former fuses to a transparent colorless glass, the latter to a grayish-green scoria or opaque glass.

Wollastonite=$3CaO, SiO^3$. The hydrochloric acid solution gives no, or only a very slight, precipitate with ammonia.

Eukolite=$SiO^3$, $NbO^3$, $Zr^2O^3$, CaO, NaO. By boiling the hydrochloric acid solution with tin, it assumes a fine blue color on reaching a certain degree of concentration. See also Wöhlerite.

Nepheline=$2 (NaO.KO), SiO^3 + 2 (Al^2O^3, SiO^3)$; Meionite=$3CaO. SiO^3+2 (Al^2O^3, SiO^3)$; Mellilite = $2(3[CaO. MgO.NaO], SiO^3) + (Al^2O^3. Fe^2O^3), SiO^3$. The hydrochloric acid solution is precipitated by ammonia. Meionite fuses with intumescence, the others quietly.

*Division* 4. *Soluble in hydrochloric acid with separation of silica, without forming a perfect jelly.* (It is sometimes necessary to treat the finely pulverized mineral with concentrated acid.)

Section 1. Giving water in a matrass.

Apophyllite $= KO.2SiO^3 + 8(CaO.SiO^3) + 16HO$; Pectolite$=3([NaO.KO].SiO^3) + 4(3CaO.2SiO^3) + 3HO$; Okenite $= 3CaO,4SiO^3 + 6HO$. The silica separates in the shape of gelatinous lumps. The hydrochloric acid solution gives no, or only a slight, precipitate with ammonia. Pectolite yields but little water, the others much. Fusibility of apophyllite=1.5, forming a white vesicular glass; of okenite=2.5—3, forming a porcelain-like mass.

Analcime$=3NaO.2SiO^3. + 3(Al^2O^3,2SiO^3)+6HO$. Gelatinizes like the preceding; in the acid solution ammonia produces a copious precipitate.

Pyrosclerite$=(Al^2O^3.Cr^2O^3), SiO^3+ 2(3[MgO.Fe]), SiO^3 + 1\frac{1}{2}HO$; Chonikrite $=2Al^2O^3, SiO^3 + 3(3[MgO.CaO.FeO], SiO^3)+6HO$, are distinguished from the other minerals of this section by their inferior hardness=2.5—3. The former gives with fluxes the reactions of oxide of chromium, §§ 67 and 68.

Brewsterite $= (SrO.BaO), SiO^3 + Al^2O^3, 3SiO^3 + 5HO$; characterized by its hydrochloric acid solution giving a precipitate with sulphuric acid.

Stilbite$=CaO.SiO^3 + Al^2O^3.3SiO^3+5HO$; Chabazite $= 3(CaO.NaO),2SiO^3 + 3(Al^2O^3.2SiO^3)+18HO$; Prehnite$=2CaO.SiO^3+Al^2O^3.SiO^3+HO$. Fuse with intumescence to enamel-like masses. Prehnite yields but little water, losing by ignition only 4.3 per cent.; the others lose from 15 to 20 per cent.

Meerschaum, see below; Deweylite=$2MgO.SiO^3+3HO$. Distinguished by being much less fusible than the preceding (fusibility=5); the former absorbs water with great avidity, the latter not.

Section 2. Giving only traces or no water in a matrass.

Tachylyte=$3(FeO.CaO.NaO), 2SiO^3+Al^2O^3, SiO^3$. Fuses readily to a black shining glass. Hardness=6.5; color black.

Scapolite=$3(CaO.NaO),SiO^3+3(Al^2O^3.SiO^3)$. Color light. Hardness=5—5.5. Fuses with intumescence to a white, vesicular glass.

Wöhlerite=$SiO^3,NbO^3,Zr^2O^3,CaO,NaO$. Fusibility =3, forming a yellowish enamel. The hydrochloric acid solution gives the same reaction as eukolite.

Labradorite=$(CaO.NaO), SiO^3+Al^2O^3.SiO^3$; Anorthite=$3CaO,SiO^3+3(Al^2O^3,SiO^3)$. Fusibility =3—4, without intumescence, forming a colorless glass; hardness of the former=6, of the latter=6—7. Cleavage perfect.

Lime Garnet (some varieties)=$3CaO, SiO^3+Al^2O^3, SiO^3$. Fusibility=3; not cleavable.

Sphene, some varieties, see below. Gives titanium-reactions, § 111.

*Division* 5. *Little affected by hydrochloric acid; give with fluxes the manganese-reactions.*

Carpholite=$SiO^3,Al^2O^3,FeO,MnO,Fe^2O^3,HO$. Occurs only in radiated and stellated tufts. Color straw-yellow; silky. Yields water.

Manganese Garnet=$3MnO.SiO^3+Al^2O^3.SiO^3$. Color brownish-red; fuses without intumescence; not cleavable.

Epidote (some varieties)=$3CaO, SiO^3+2([Al^2O^3.Mn^2O^3.Fe^2O^3],SiO^3)$. Fusibility=2—2.5, intumesces. Cleavable. Color cherry-red to reddish-black.

Rhodonite=$3MnO,2SiO^3$. Fusibility=3, without intumescence. Color rose-red; cleavable.

*Division* 6. *Not belonging to either of the preceding divisions.*

Scheelite=$CaO,WO^3$. Fusibility=5. Soluble in hydrochloric acid, leaving a residue of tungstic acid, which is soluble in ammonia, and which gives with SPh the characteristic reaction of tungstic acid; see Table II.

Lepidolite=$KO, LiO, F, Al^2O^3, SiO^3$; Euphyllite=$SiO^3, Al^2O^3, CaO, HO$; Margarite=$SiO^3, Al^2O^3, CaO, HO$. Fusibility of lepidolite=2; gives the lithia-reaction, § 89. Fusibility of euphyllite=4.5, of margarite=4. Color of euphyllite white to colorless; of margarite grayish, reddish-white, yellowish. All three possess perfect cleavage.

Petalite=$3(LiO.NaO), 2SiO^3 + 4(Al^2O^3, 3SiO^3)$; Spodumene=$3(LiO.NaO), 2SiO^3 + 4(Al^2O^3, 2SiO^3)$, do not possess as perfect a cleavage as the preceding, and greater hardness; hardness of petalite=6—6.5, of spodumene=6.5—7. Both give the lithia-reaction, § 89. Spodumene fuses with intumescence to a glassy globule; petalite fuses to a white enamel.

Diallage=$3(CaO.MgO), 2SiO^3$. Fusibility=3.5; characterized by its pearly metallic lustre; cleaves easily in one direction.

Harmotome=$BaO,SiO^3+Al^2O^3,2SiO^3+5HO$. Distinguished from the other minerals of this division by yielding water in a matrass. Occurs usually in twin crystals.

Axinite=$SiO^3,Al^2O^3,CaO,FeO,MnO,BO^3$; Tourmaline=$SiO^3,Al^2O^3,FeO,KO,NaO,LiO,BO^3$. Give the reaction of boracic acid, § 61. Axinite fuses readily with intumescence to a dark-green glass. Different varieties of tourmaline show different fusibility. Hardness of axinite=6.5—, of tourmaline=7—7.5.

Diopside (white augite)=$3CaO, 2SiO^3 + 3MgO, 2SiO^3$; Augite=$3CaO, 2SiO^3 + 3(MgO.FeO)$,

$2SiO^3$. Hardness=6; diopside fuses to a whitish, augite to a black glass. Color of augite black or dark-green; of diopside pale-green or gray, or colorless.

Tremolite = $CaO.SiO^3 + 3MgO.2SiO^3$; Hornblende = $CaO.SiO^3 + 3(MgO.FeO), 2SiO^3$. Hardness= 5.5; fusibility=3—4. Tremolite fuses to a white or light-colored glass, hornblende to a black or gray glass; the former is colorless or white, or of light green, yellow, or gray color; hornblende is green or black.

Sphene=$2(CaO,SiO^3) + CaO,3TiO^2$. Fusibility=3. Hardness=5—5.5. Gives the titanium-reaction, § 111. Imperfectly soluble in hydrochloric acid.

Orthoclase=$KO.SiO^3 + Al^2O^3,3SiO^3$; Albite=$NaO, SiO^3 + Al^2O^3,3SiO^3$. Hardness=6. Fuse without intumescence; fusibility of orthoclase=5, of albite=4; the latter colors the flame yellow. Not soluble in acids. With solution of cobalt become blue on the edges, § 44.

Zoisite=$3CaO.SiO^3 + 2(Al^2O^3,SiO^3)$;Epidote=$CaO, SiO^3 + 2([Al^2O^3.Fe^2O^3], SiO^3)$. Hardness = 6.5. Fusibility = 3—3.5; fuse with intumescence, zoisite to a white or yellowish slag, epidote to a black or dark-brown slag. Color of zoisite gray, yellowish-gray, grayish-white; of epidote green.

Lime Garnet=$3CaO,SiO^3 + Al^2O^3,SiO^3$; Idocrase= $3CaO,SiO^3 + (Fe^2O^3.Al^2O^3,SiO^3$; Pyrope=$(MgO.FeO.CaO),SiO^3 + (Al^2O^3.Cr^2O^3), SiO^3$. Hardness =6.5—7.5. Fusibility of lime garnet and idocrase =3,of pyrope=4.5. Idocrase possesses cleavage, the others not. Pyrope gives with the fluxes the chromium-reactions.

(See, also, emerald, euclase, iolite, biotite, and muscovite.)

Obsidian, Pitchstone, Pearlstone, and Pumice= $SiO^3,Al^2O^3,NaO,KO,HO$, are amorphous. Fusibility=3.5—4, fuse with intumescence to porcelain-like masses, or white vesicular glasses. Lustre of obsidian glassy, of pitchstone greasy, of pearlstone pearly; pumice is characterized by its porosity.

CLASS III. INFUSIBLE, OR FUSIBILITY ABOVE 5.

*Divisiou* 1. *After ignition moistened with solution of cobalt and again ignited, assume a bright-blue color.*

With the hard, anhydrous minerals of this division, the color is best seen by reducing the substance to a fine powder and moistening this with the solution of cobalt. The color appears only after cooling.

Section 1. Giving much water in a matrass.

Alunite=$SO^3, Al^2O^3$, KO, HO; Websterite=$Al^2O^3$, $SiO^3+9HO$. Give a sulphur-reaction, § 107. Websterite is readily soluble in hydrochloric acid; alunite not visibly affected.

(See, also, ammonia alum, and potash alum.)

Plumbo-Resinite, see § 176.

Calamine, see § 214.

Wavellite = $4Al^2O^3, 3PO^5 + 18HO$; Gibbsite = $Al^2O^3, PO^5 + 8HO$; Peganite= $2Al^2O^3, PO^5 + 6HO$; Fischerite=$2Al^2O^3, PO^5+2HO$. Soluble to a great extent in hydrate of potassa. Give the reactions of phosphoric acid, §§ 94 and 95. The former two occur usually in globular concretions of radiated structure, the latter two minutely crystalline. Peganite loses on ignition 24 per cent. of water, wavellite 27, fischerite 29, gibbsite, 35.

Diaspore = $Al^2O^3$, HO; Clintonite = $SiO^3$, $Al^2O^3$, CaO, MgO, HO. Diaspore is but slightly soluble in hydrate of potassa; clintonite insoluble; the former loses on ignition $11\frac{1}{2}$ per cent. of water, the latter $4\frac{1}{2}$. Hardness of diaspore= 6.5—7; of clintonite=4—5.

Allophane=3 $Al^2O^3$, 2 $SiO^3$+15 HO; Halloysite =3 $Al^2O^3$, 4$SiO^3$+12HO; Ochran=$Al^2O^3$, $SiO^3$ +6HO; Collyrite=3$Al^2O^3$, $SiO^3$+15HO. Decomposed by hydrochloric acid with separation of gelatinous silica. Hardness of allophane=3, of the others=1—2. Halloysite loses on ignition 16 per cent. of water, ochran 21, collyrite 33½.

Pholerite=$Al^2O^3$, $SiO^3$+2HO; Cimolite=$Al^2O^3$, 3$SiO^3$+3HO; Kaolin=3$Al^2O^3$, 4$SiO^3$+6HO and 2$Al^2O^3$,3$SiO^3$+6HO; are all very soft and earthy, and but little affected by acids; lose on ignition from 12 to 16 per cent. of water. Nearly related to these minerals are the various varieties of common clay, some varieties of lithomarge (with 14 per cent. of water), and bole with 24—26 per cent. of water; the clays become plastic with water, the latter two not.

Section 2. Giving little or no water in a matrass.

Lazulite=$PO^5$, $Al^2O^3$, MgO, FeO, HO. Gives the reaction of phosphoric acid, § 74. Heated, loses its blue color and becomes white. Not affected by acids.

Willemite=3ZnO, $SiO^3$. With solution of cobalt (§ 44) becomes blue, and green in spots. Gelatinizes with hydrochloric acid.

Myelin=2($Al^2O^3$, $SiO^3$)+HO; Agalmatolite=$SiO^3$, $Al^2O^3$, KO, HO; Pyrophyllite=3MgO, 2$SiO^3$+ 9($Al^2O^3$, $SiO^3$)+9HO. Are very soft, hardness= 1—2. Pyrophyllite is foliated like talc; before the Blp swells up and spreads out into fan-like shapes, increasing to about 20 times its former bulk. The others do not change before the Blp. Myelin is partially decomposed by hydrochloric acid; agalmatolite not affected.

Muscovite=KO, $SiO^3$+4($Al^2O^3$, $SiO^3$). Cleavage eminent in one direction; folia elastic. Does not swell perceptibly before the Blp, fusible in very thin laminæ. Not affected by acids. Hardness=2.5.

Disterrite (variety of clintonite), cleavable in one direction. Hardness=4—5. Decomposed by concentrated sulphuric acid.

Andalusite=$4Al^2O^3,3SiO^3$; Kyanite=$3Al^2O^3,2SiO^3$; are but little affected by acids. Kyanite occurs generally in bladed crystallizations; hardness=6—7. Hardness of andalusite=7.5, but variety chiastolite varies in hardness from 3 to 7.5.

Topaz=$2Al^2F^3+5(Al^2O^3,SiO^3)$; Lithia Tourmaline =$SiO^3$, $BO^3$, $Al^2O^3$, MnO, LiO, KO. Not affected by acids. Not completely soluble in SPh, the glass becomes opalescent on cooling. Topaz on being ignited remains transparent and does not swell; tourmaline becomes white and swells. Topaz is cleavable in one direction. Hardness =8; tourmaline is not cleavable, hardness=6.5.

Corundum (sapphire)=$Al^2O^3$; Chrysoberyl=BeO, $Al^2O^3$. Not affected by acids. When pulverized, slowly but completely soluble in SPh; the glass does not opalesce on cooling. Hardness of chrysoberyl=8.5, of corundum=9; color of the former usually green, of the latter blue, red, yellow, brown.

(Some varieties of Spinel and Leucite assume a blue color with solution of cobalt.)

*Division* 2. *Moistened with solution of cobalt and ignited, assume a green color.*

It is sufficient to heat to redness. The minerals of this division give a coating of oxide of zinc on charcoal, § 25.

Smithsonite, see § 213.

Zinc Bloom=$(ZnO,CO^2+HO)+2(ZnO,HO)$. Dissolves readily in hydrochloric acid with effervescence; the solution gives with ammonia a white precipitate, soluble in an excess of the reagent. Yields water in a matrass.

Willemite$=3ZnO,SiO^3$; Calamine, see § 214. Gelatinize with hydrochloric acid. Calamine yields water, willemite not. With solution of cobalt assume a green color only in spots.

(See also Blende and Goslarite.)

*Division* 3. *After ignition have an alkaline reaction, and change into blue the color of a moistened red litmus-paper.*

Brucite$=MgO, HO$; Hydromagnesite$=MgO,4HO+3(MgO,CO^2)$. Yield much water in a matrass, unlike the other minerals of this division. Brucite dissolves in hydrochloric acid without effervescence, hydromagnesite with effervescence; the concentrated solutions are not precipitated by sulphuric acid. Lancasterite is a mixture of brucite and hydromagnesite. Nemalite is a fibrous variety of brucite, of silky lustre.

Calcite$=CaO. CO^2$; Arragonite$=CaO. CO^2$. Dissolve readily and with effervescence in dilute cold hydrochloric acid; the concentrated (but not the dilute) solution gives a precipitate with sulphuric acid. Arragonite falls to powder before the Blp, calcite not.

Dolomite $= MgO. CO^2+CaO. CO^2$; Magnesite$=MgO. CO^2$. Do not, or but slightly, effervesce with cold dilute hydrochloric acid, but dissolve readily on application of heat. The concentrated solution of the former gives a precipitate with sulphuric acid, that of the latter not.

A similar behavior shows the Breunnerite$=(MgO. FeO. MnO), CO^2$, which on ignition becomes black and slightly magnetic; and some varieties of Chalybite, see § 163, and Diallogite, see § 185.

Strontianite$=SrO.CO^2$; Barytocalcite$=BaO.CO^2+CaO.CO^2$. Dissolve with effervescence in dilute hydrochloric acid; the solution, even if largely diluted with water, gives a precipitate with sulphuric acid. Strontianite colors the flame red, § 34; barytocalcite yellowish-green, § 35.

(See also Yttrocerite.)

*Division 4. Completely soluble, or nearly so, in hydrochloric or nitric acid without gelatinizing or leaving a perceptible residue of silica.*

Chalybite, see § 163; Breunnerite, see preceding division; Diallogite, see § 185; Emerald Nickel, see § 195. Dissolve in heated hydrochloric acid with effervescence.

Limonite, see § 155; Göthite=$Fe^2O^3,HO$. Become black and magnetic in reduction flame. Dissolve in hydrochloric acid without effervescence. Göthite occurs crystallized and cleaves distinctly in one direction; loses 10 per cent. on ignition; limonite loses 14½ per cent.

(See also Hematite which in some varieties is without metallic lustre; readily distinguished by red streak.)

Blende, see § 212; Marmatite=$FeS+3ZnS$; Greenockite=$CdS$. Dissolve in hydrochloric acid with evolution of sulphuretted hydrogen. Give the sulphur-reaction, § 107. Greenockite gives on charcoal a coating of oxide of cadmium, § 24, the others of oxide of zinc, § 25. Marmatite gives after calcination with the fluxes the reactions of iron.

Wad, see § 184; Zincite, see § 211.

Earthy Cobalt, see § 133. Some varieties are fusible.

Pitchblende = $UO,U^2O^3$; Zippeite = $U^2O^3+xHO$. Give with the fluxes the reactions of sesquioxide of uranium [Table II]. Give with nitric acid a yellow solution in which ammonia produces a sulphur-yellow precipitate. Pitchblende is black, zippeite yellow.

Chrome Ochre=$Cr^2O^3$. Gives with fluxes the

reactions of sesquioxide of chromium [Table II]. Forms with hydrate of potassa a green solution.

Turquois = $PO^5, Al^2O^3, HO, CuO$. Color sky-blue and green. Gives the copper-reaction, § 74. Yields much water in a matrass.

Apatite = $3(3CaO. PO^5) + Ca(Cl, F)$. Gives the phosphoric acid reaction, § 94. Fusibility = 5. Soluble in nitric acid. Gives the fluorine-reaction, § 76 (always ?)

Monazite = $PO^5, CeO, LaO, ThO$. Infusible. Gives the phosphoric acid reaction, § 94. Soluble in hydrochloric acid. Minute tabular crystals of reddish-brown color.

Childrenite = $PO^5, Al^2O^3, FeO, MnO, HO$. Gives the phosphoric acid reaction, § 94. With the fluxes gives the reaction of iron and manganese. In hydrochloric acid soluble with difficulty. Yields much water.

Polycrase = $TiO^2, NbO^3, Zr^2O^3, Fe^2O^3, Ce^2O^3, U^2O^3$, &c. Decrepitates, but infusible. Color black. On fusing the pulverized mineral with hydrate of potassa, boiling the fused mass with hydrochloric acid, filtering, and boiling the filtrate with tin-foil, the liquid assumes a blue color on reaching a certain degree of concentration; the color disappears on addition of water.

Fluocerite = $CeF$. Gives the reactions of fluorine, § 75, and of sesquioxide of cerium, Table II. Yttrocerite = $F, CaO, YO, Ce^2O^3$, behaves similarly.

*Division* 5. *With hydrochloric acid form a jelly, or are decomposed with separation of silica without gelatinizing.*

Section 1. Giving water in a matrass.

Dioptase = $3CuO, 2SiO^3 + 3HO$; Chrysocolla, see § 149. Behave alike before the Blp; the former gelatinizes with acids, the latter not.

Thorite = $3ThO, SiO^3 + 3HO$; Cerite = $3CeO, SiO^3 +$

3HO. Gelatinize with hydrochloric acid. Color of thorite orange-yellow or black, hardness= 4.5—5; of cerite, brown to red passing into gray, hardness=5.5.

Chloropal=$Fe^2O^3,2SiO^3+3HO$. Color yellowish-green, amorphous, of an opal-like appearance. Becomes magnetic by ignition; gelatinizes. Small pieces when thrown into a concentrated solution of hydrate of potassa lose the green color and become dark-brown. Hardness=2—3.

Hisingerite=$Fe^2O^3,SiO^3+3HO$; Xylotile=$Fe^2O^3,3SiO^3+3MgO,2SiO^3+5HO$. Become magnetic by ignition. Readily decomposed by hydrochloric acid. Hisingerite is black, imperfectly crystallized and cleavable in one direction; xylotile is light or dark brown, of fibrous, woody structure.

Meerschaum=$MgO,SiO^3+2HO$. Gelatinizes with hydrochloric acid; very light; absorbs water with great avidity; gives the magnesia-reaction with solution of cobalt, § 44.

Schiller-Spar = $3([MgO. FeO], SiO^3) + 2(MgO, 2HO)$; Chrysotile = $3MgO,2SiO^3 + MgO,3HO$. Possess a metallic pearly lustre; the former is massive, cleavable; the latter fibrous. By ignition schiller-spar becomes brown, chrysotile white. Both are decomposed by hydrochloric acid, or more readily by sulphuric acid, without gelatinizing.

Serpentine=$2(3MgO,2SiO^3)\times 3(MgO,2HO)$. Decomposed by concentrated hydrochloric acid without gelatinizing. Usually massive and compact; hardness=3—4; loss by ignition 12—13 per cent. Of similar composition, and showing a similar behavior, are the following minerals, which, however, possess crystalline structure and cleavage: Picrophyll, hardness=2.5, loss by ignition $10\frac{1}{2}$ per cent.; Picrosmine, hardness=2.7, loss by ignition 9 per cent.; Marmolite, hardness =2.5—3, loss by ignition 15.7 per cent.; Kæmmererite, hardness=1.5—2, loss by ignition 13 per cent.

(See also Chlorite and Ripidolite which are with difficulty decomposed by concentrated hydrochloric acid.)

Antigorite=$3(MgO.FeO),2SiO^3+MgO,HO$; Monradite=$4(3[MgO.FeO],2Sio^3)+3HO$; Neolite=$3MgO,2SiO^3+HO$. Decomposable by concentrated hydrochloric acid without gelatinizing. Loss by ignition 4—6 per cent. Antigorite occurs in foliated masses, hardness=2.5; monradite, hardness=6; neolite in silky fibres or massive, hardness=1.

(See, also, some varieties of Clintonite, hardness= 4—5.)

Section 2. Giving only traces or no water in a matrass.

Gadolinite=$SiO^3,YO,FeO,CeO,Be^2O^3$; Gehlenite=$2(3CaO,SiO^3)+2(Al^2O^3.Fe^2O^3),SiO^3$. Gelatinize with hydrochloric acid. Gadolinite swells before the Blp into cauliflower-like masses, and sometimes exhibits a vivid glow; thin splinters fusible on the edges; color black to blackish-green; hardness=6.5—7. Gehlenite is also fusible in very thin splinters; color gray to grayish-white; hardness=5.5—6.

Chrysolite=$3MgO,SiO^3$; Chondrodite=$2(3MgO,SiO^3)+MgF$. Gelatinize with hydrochloric acid. Color of the former green, of the latter mostly white, yellow, or brown. Chondrodite gives the fluorine-reaction, § 76.

Boltonite (a variety of pyroxene)=$SiO^3,MgO,FeO,Al^2O^3$. Cleavage distinct in one direction. Color yellow. (See also Clintonite.)

Leucite = $3KO, 2SiO^3 + 3(Al^2O^3,2SiO^3)$. Decomposed by hydrochloric acid, the silica separating as a fine powder; some varieties become blue with solution of cobalt; occurs usually in trapezohedrons. Color grayish or white.

*Division* 6. *Not belonging to either of the preceding divisions.*

The remaining minerals which cannot be classed under any of the preceding divisions, may be divided according to their hardness in two sections.

Section 1. Hardness below 7.

Biotite (hexagonal mica)=$(Al^2O^3.Fe^2O^3),SiO^3+3(MgO,KO),SiO^3$; Muscovite (oblique mica)= $KO,SiO^3+4(Al^2O^3,SiO^3)$; Talc=$6MgO,5SiO^3+2HO$. Give little or no water in a matrass; talc loses at most 5 per cent. Cleavage eminent in one direction. Hardness of biotite=2.5—3, of muscovite=2—2.5, of talc=1—1.5. Biotite is decomposed by concentrated sulphuric acid, the others not. The laminæ of biotite and muscovite are elastic, of talc not. Soapstone or steatite is a massive, usually compact variety of talc; very greasy to the feel, or like soap (see also Pyrophyllite).

Chlorite = $2(MgO,Al^2O^3)+3(2[MgO.Fe],SiO^3)+6HO$; Ripidolite = $(MgO.FeO),SiO^3+(Al^2O^3.Fe^2O^3),SiO^3+4(MgO,HO$. Loses by ignition 12 per cent. of water. Cleavage eminent in one direction, laminæ not elastic (chlorite often massive granular). Hardness of chlorite=2—2.5, of ripidolite=1—2. Decomposed by concentrated hydrochloric acid with continued boiling, more readily by sulphuric acid. Ripidolite fuses with difficulty (=5.5) to a grayish-yellow enamel, chlorite becomes black and slightly magnetic. A similar behavior shows the Chloritoid; hardness=5.5—6.

Bronzite (hypersthene) = $3MgO, 2SiO^3+3(CaO$.

FeO), $2SiO^3$; Anthophyllite=$FeO,SiO^3+3MgO,2SiO^3$. Cleavage of bronzite very perfect in one direction; anthophyllite cleaves in two directions. The former is of clove-brown or pinchbeck-brown color, with a pearly-metallic lustre; the lustre of anthophyllite is much less perfect. Hardness=5—5.5.

Wolframite=$WO^3$, is soluble in hydrate of potassa; the solution gives with nitric acid a yellow precipitate which on boiling becomes lemon-yellow. Occurs in soft, earthy, yellow masses.

Scheelite=$CaO,WO^3$. Fusibility=5; hardness=4.5—5. The pulverized mineral, on being boiled with nitric acid, leaves a lemon-yellow residue of tungstic acid. Gives the reactions of tungstic acid [Table II].

Cassiterite, see § 209.

Anatase, Rutile, and Brookite=$TiO^2$. Give the reactions of titanic acid [Table II]. On fusing the pulverized minerals with hydrate of potassa, dissolving the fused mass in hydrochloric acid and boiling the solution with metallic tin, it assumes a violet color, which turns to red on addition of water. Color of anatase, various shades of brown passing into indigo-blue; of rutile mostly brownish-red or red, sometimes yellowish or black; of brookite, hair-brown, yellowish or reddish (variety arkansite is iron-black). Hardness of anatase=5.5—6; of rutile=6—6.5; of brookite=5.5—6.

Aeschynite and Pyrochlore=$NbO^3,TiO^2,Zr^2O^3,Ce^2O^3,CaO$, &c. Treated like the preceding with potassa, &c., the solution on reaching a certain degree of concentration assumes a fine blue color, which, on addition of water, does not change to red, but gradually disappears. Aeschynite swells before the Blp and turns yellow; pyrochlore does not swell and becomes grayish.

Opal=$SiO^3$+xHO. Before the Blp yields water and becomes opaque; fuses with carbonate of soda to a clear bead, with effervescence. Infusible. Boiled with hydrate of potassa, it dissolves completely or to a great extent; the solution gives a gelatinous precipitate with chloride of ammonium. Hardness=6—6.5.

Xenotime=3YO,$PO^5$. Color, various shades of brown or flesh-red. Hardness=4—5. Gives the phosphoric acid reaction, § 94. Infusible. With salt of phosphorus dissolves with great difficulty to a colorless glass.

[See also Childrenite and Orthoclase.]

Section 2. Hardness=7, or above.

[See Cassiterite, Rutile, and Opal of the preceding section whose hardness sometimes approaches 7.]

Quartz=$SiO^3$. The various varieties of quartz, as rock-crystal, amethyst, hornstone, flint, chalcedony, &c., are infusible and unalterable before the Blp, and fuse with carbonate of soda to a transparent bead, with effervescence. Hardness=7.

Iolite=2 (3[MgO.FeO], $2SiO^3$) + 5($Al^2O^3$, $SiO^3$); Staurotide=2($Al^2O^3$.$Fe^2O^3$), $SiO^3$. Hardness=7—7.5. Do not fuse to a transparent glass with carbonate of soda. Fusibility of iolite=5—5.5; color blue, grayish. Staurotide is infusible; color brownish-red, brown; crystals often cruciform.

Beryl = 3BeO, $2SiO^3$ + $Al^2O^3$, $2SiO^3$; Euclase = 2 (3BeO, $SiO^3$) + $2Al^2O^3$, $SiO^3$; Phenacite=3BeO, $SiO^3$; Zircon=$Zr^2O^3$,$SiO^3$. Hardness=7.5. Beryl and euclase turn milk-white with strong heat and become rounded on the edges; beryl crystallizes in hexagonal prisms, and possesses pretty distinct basal cleavage, color usually pale-green or emerald-green; euclase crystallizes in clinorhombic prisms and possesses distinct cleavage in two directions at right angles to each other; color pale mountain-green passing into blue and white. Phenacite and zircon do no change before the blow-pipe, excepting that zircon becomes colorless; color red, yellow, or colorless, zircon sometimes brown or gray; phenacite is a little harder (=8) than zircon.

Ouvarovite (lime-chrome-garnet) = $3CaO, SiO^3 + Cr^2O^3, SiO^3$. Infusible. Hardness = 7.5—8. Gives with fluxes the chromium reactions [Table II].

Spinel = $MgO, Al^2O^3$; Pleonaste = $(MgO.FeO), Al^2O^3$; Gahnite = $(ZnO.MgO), SiO^3$. Hardness = 7.5—8. Occur almost exclusively in octahedral crystals. Spinel and pleonaste, when pulverized, are soluble in salt of phosphorus; color of spinel red, blue, brownish; of pleonaste black. Gahnite is almost insoluble in salt of phosphorus and borax; color dark-green or black. Kreittonite is a black spinel containing zinc and iron, slightly magnetic before ignition.

Diamond = C. Characterized by its hardness, which surpasses that of corundum.

# TABLES.

## TABLE I.—BEHAVIOR OF THE ALKALINE BEFORE THE

| | *On Ch alone, and in the forceps.* | *With Carbonate of Soda on Ch.* |
|---|---|---|
| 1. BARYTA. BaO. | The Hydrate fuses, boils, intumesces, and is finally absorbed by the Ch. The Carbonate fuses readily to a transparent glass, which, on cooling, becomes enamel-white. In the forceps it colors the outer flame yellowish-green. | Fuses with Sd to a homogeneous mass, which is absorbed by the Ch. |
| 2. STRONTIA. SrO. | The Hydrate behaves like hydrate of Baryta. The Carbonate fuses only at the edges, and swells out in arborescent ramifications which emit a brilliant light, and, when heated with the RFl, impart to it a reddish tinge; shows after cooling alkaline reaction. In the forceps, colors the outer flame purple. | Caustic Strontia is insoluble. The Carbonate, mixed with its own volume of Sd, fuses into a limpid glass, which becomes enamel-white on cooling. At a greater heat the mass enters into ebullition, and caustic Strontia is formed, which is absorbed by the Ch. |
| 3. LIME. CaO. | Caustic Lime suffers no alteration. The Carbonate loses carbonic acid, becomes whiter and more luminous, and shows after cooling alkaline reaction. In the forceps it colors the outer flame pale-red. | Insoluble. The Sd passes into the Ch, and leaves the Lime unaltered on its surface. |
| 4. MAGNESIA. MgO. | Undergoes no alterations. The Carbonate becomes caustic and luminous. | It behaves like Lime. |
| 5. ALUMINA. $Al^2O^3$. | Not changed. | Forms an infusible compound, with slight intumescence. The excess of Sd is absorbed by the Ch. |

EARTHS AND THE EARTHS PROPER
BLOW-PIPE.

| *With Bx on Platinum Wire.* | *With SPh on Platinum Wire.* |
| --- | --- |
| The Carbonate dissolves with effervescence to a limpid glass which, when in a certain state of saturation, may be made opaque by flaming; when still more saturated, it becomes opaque on cooling, even without flaming. | As with Borax. |
| Presents the same phenomena as Baryta. | Presents the same phenomena as Baryta. |
| Readily dissolved to a limpid glass, which becomes opaque by flaming. The Carbonate dissolves with effervescence. On a large addition of Lime the glass crystallizes on cooling, but does not become enamel-white. | Soluble in large quantities to a limpid glass which, when sufficient Lime is present, becomes opaque by flaming. When saturated, the glass becomes enamel-white on cooling. |
| It behaves like Lime, but does not crystallize so well. | Readily soluble to a limpid glass, which becomes opaque by flaming. When saturated, it becomes, on cooling, enamel-white. |
| Dissolves slowly to a limpid glass, which remains so on cooling, and which cannot be made cloudy by flaming. A large quantity of Alumina makes the glass cloudy; on cooling, it then assumes a crystalline surface. | Soluble to a limpid glass, which remains clear under all circumstances. If too much Alumina is added, the undissolved portion becomes translucent. |

TABLE I.—CON-

| | *On Ch alone, and in the forceps.* | *With Carbonate of Soda on Ch.* |
|---|---|---|
| 6. GLUCINA. BeO. | Not changed. | Insoluble. |
| 7. YTTRIA. YO. | Not changed. | Insoluble. |
| 8. ZIRCONIA. $Zr^2O^3$. | Infusible, but emitting a very glaring light. | Insoluble. |

TINUED.

| *With Bx on Platinum Wire.* | *With SPh on Platinum Wire.* |
|---|---|
| Soluble in large quantities to a limpid glass, which becomes opaque by flaming. When Glucina is present in excess, it becomes enamel-white on cooling. | As with Borax. |
| Like Glucina. | Like Glucina. |
| Like Glucina. | Dissolves more slowly than with Borax. |

TABLE II.—BEHAVIOR OF THE METAL-

| *Metallic Oxides in Alphabetical Order.* | *On Charcoal alone.* | *With Carbonate of Soda.* |
|---|---|---|
| 1. ANTIMONIOUS ACID. $SbO^3$. | OFl: It is displaced without change, and deposited upon another part of the Ch.<br>RFl: It is reduced and volatilized. A Ct of antimonious acid is deposited on the Ch, and a greenish-blue color imparted to the flame. | On Ch very readily reduced in OFl and RFl. The metal fumes and coats the Ch with antimonious acid. |
| 2. ARSENOUS ACID. $AsO^3$. | Volatile below red heat. | On Ch reduced, with emission of arsenical fumes, which are characterized by a strong garlic odor. |
| 3. TEROXIDE OF BISMUTH. $BiO^3$. | OFl: On platinum foil it fuses readily to a dark-brown mass, which, on cooling, becomes pale yellow.<br>On Ch in OFl and RFl reduced to metallic bismuth, which, with long blowing, vaporizes, coating the Ch with yellow oxide. The Ct, when touched with the RFl, disappears without coloring the flame. | Easily reduced to metallic bismuth. |
| 4. OXIDE OF CADMIUM. CdO. | OFl: On platinum foil unchanged.<br>RFl: On Ch it disappears in a short time, and deposits all over the Ch a dark-yellow or reddish-brown powder; the color can only be clearly discerned after cooling. | OFl: Insoluble.<br>RFl: On Ch readily reduced; the metal vaporizes and deposits a dark-yellow or reddish-brown Ct on the Ch. |

LIC OXIDES BEFORE THE BLOW-PIPE.

| *With Bx on Platinum Wire.* | *With SPh on Platinum Wire.* |
| --- | --- |
| OFl: Dissolves in large quantities to a limpid glass, which, while hot, appears yellowish, but after cooling, colorless. RFl: The glass, when treated only for a short time in the OFl, becomes on Ch grayish and cloudy from particles of reduced antimony. With tin it becomes gray or black. | OH: Dissolves with effervescence to a limpid glass, which, while hot, is slightly yellowish. RFl: On Ch the saturated bead becomes at first cloudy, but afterwards clear again, owing to the volatilization of the reduced antimony. Treated with tin, the glass becomes, after cooling, gray, even if but very little antimonious acid is present. With strong blowing it becomes clear again. |
| 0 | 0 |
| OFl: A small quantity is easily dissolved to a clear yellow glass, which, on cooling, becomes colorless. On a large addition of oxide, the glass, while hot, is yellowish-red, becomes yellow on cooling, and when cold is opalescent. RFl: On Ch the glass becomes at first gray and cloudy, the oxide is reduced to metal with effervescence, and the bead becomes clear again. An addition of tin accelerates the process. | OFl: Readily dissolved to a limpid yellow glass, which, on cooling, becomes colorless. When a greater quantity of oxide is present, the glass may be made enamel-white by flaming, and on a still larger addition it becomes by itself enamel-white on cooling. RFl: On Ch, particularly when tin is added, the glass remains colorless and limpid while hot, but becomes, on cooling, dark-gray and opaque. |
| OFl: Soluble in large quantity to a limpid yellowish glass, becoming almost colorless on cooling. When highly saturated, it may be made enamel-white by flaming, and when still more oxide is present, it becomes by itself enamel-white on cooling. RFl: Placed on Ch, it enters into ebullition; the oxide is reduced; the reduced metal vaporizes immediately and deposits a dark-yellow Ct. | OFl: Soluble in large quantity to a limpid glass which, while hot, is yellowish, but colorless when cold; when saturated, it becomes enamel-white on cooling. RFl: On Ch the oxide becomes slowly and imperfectly reduced. The reduced metal deposits a very feeble Ct of dark-yellow color. The color is only clearly seen when the mass is cold. An addition of tin facilitates the reduction. |

TABLE II. — CON-

| *Metallic Oxides in Alphabetical Order.* | *On Charcoal alone.* | *With Carbonate of Soda.* |
|---|---|---|
| 5. SESQUIOXIDE OF CERIUM. $Ce^2O^3$. | Not changed. | Insoluble. The Sd passes into the Ch; the sesquioxide is reduced to protoxide, which remains on the Ch as a light-gray powder. |
| 6. SESQUIOXIDE OF CHROMIUM. $Cr^2O^3$. | Not changed. | OFl: On platinum wire soluble to a dark yellowish-brown glass, which on cooling becomes opaque and yellow. RFl: The glass becomes opaque and green on cooling. On Ch it cannot be reduced to metal; the Sd passes into the Ch, and the oxide remains behind as a green powder. |
| 7. OXIDE OF COBALT. CoO. | OFl: Not changed. RFl: It is reduced to metal, but does not fuse: the mass is attracted by the magnet, and assumes metallic lustre by friction. | OFl: On platinum wire a very small quantity is dissolved to a transparent mass of a pale-reddish color, which on cooling becomes gray. RFl: On Ch reduced to a gray magnetic powder. |
| 8. OXIDE OF COPPER. CuO. | OFl: Fuses to a black globule, which becomes reduced when it is in contact with the Ch. RFl: Reduced to metal at a temperature below the melting-point of copper. When the heat is increased a globule of metallic copper is obtained. | OFl: On platinum wire soluble to a limpid glass of green color; on cooling it becomes opaque and white. RFl: On Ch easily reduced to metal, which, when the temperature is sufficiently high, fuses to one or more globules. |

TINUED.

| With Bx on Platinum Wire. | With SPh on Platinum Wire. |
|---|---|
| OFl: Soluble to a limpid glass of dark-yellow or red color, which changes on cooling to yellow. When highly saturated with oxide the glass becomes on cooling enamel-white. RFl: The yellow glass becomes colorless. A highly saturated bead becomes on cooling enamel-white and crystalline. | OFl: As with Bx, but on cooling colorless. RFl: Perfectly colorless, hot and cold. Becomes never opaque on cooling, however large the amount of oxide. |
| OFl: Dissolves but slowly, but colors intensively. If little of the oxide is present, the glass, while hot, is yellow; when cold, yellowish-green; with more oxide it is dark-red while hot, becomes yellow on cooling, and when perfectly cold has a fine yellowish-green color. RFl: The glass is green, hot and cold. The intensity of the color depends on the amount of oxide present. Tin causes no change. | OFl: Soluble to a limpid glass, which, while hot, appears reddish; when cold it has a fine green color. RFl: As in OFl. |
| OFl: Colors very intensively. The glass appears pure smalt-blue, hot and cold. An excess of oxide imparts to the bead a deep bluish-black color. RFl: As in OFl. | OFl: As with Bx, but for the same quantity of oxide the color is not quite so deep. RFl: As in OFl. |
| OFl: A small addition of oxide makes the glass appear green while hot, but blue when cold. A large quantity imparts to it a very deep-green color while hot, becoming greenish-blue when cold. RFl: A glass containing a certain quantity of oxide becomes colorless, but on cooling becomes opaque and red (suboxide). On Ch the copper may be precipitated in the metallic state, the bead becoming in consequence colorless. A glass containing protoxide, when treated on Ch with tin, becomes on cooling brownish-red and opaque. | OFl: As with Bx, but for the same amount of oxide the coloration is not so deep. RFl: A glass containing a large quantity of oxide becomes dark-green, which in the moment of refrigeration changes suddenly to brownish-red and opaque. A glass containing but little oxide, when treated on Ch with tin, appears colorless while hot, but becomes brownish-red and opaque on cooling. |

TABLE II. — CON-

| *Metallic Oxides in Alphabetical Order.* | *On Charcoal alone.* | *With Carbonate of Soda.* |
|---|---|---|
| 9. TEROXIDE OF GOLD. $AuO^3$. | When heated to ignition it becomes reduced to metal in OFl and RFl. The metal fuses easily to a globule. | Does not dissolve in the Sd, but is easily reduced in both flames. The metal fuses readily to a globule. The Sd passes into the Ch. |
| 10. SESQUIOXIDE OF IRON. $Fe^2O^3$. | OFl: Not changed. RFl: Becomes black and magnetic. | OFl: Insoluble. RFl: On Ch it is reduced; the mass, when placed in a mortar, pulverized, and repeatedly washed with water to remove the adherent Ch particles, yields a gray metallic powder which is attracted by the magnet. |
| 11. BINOXIDE OF IRIDIUM. $IrO^2$. | At a red heat becomes reduced; the reduced metal is infusible. | OFl: Does not dissolve in the Sd, but becomes reduced; the metal cannot be fused to a globule. RFl: As in OFl. |
| 12. OXIDE OF LEAD. PbO. | Minium, when heated on platinum foil, blackens; on increasing the temperature it changes into yellow oxide, which finally fuses to a yellow glass. On Ch in OFl and RFl almost instantaneously reduced to metal which, with continued blowing, vaporizes, and covers the Ch with yellow oxide, surrounded by a faint white ring of carbonate. The Ct, when touched with the RFl, disappears, imparting to the flame an azure-blue tinge. | OFl: On platinum wire readily dissolved to a limpid glass which, on cooling, becomes yellowish and opaque. RFl: On Ch reduced to metal which, with continued blowing, covers the Ch with oxide. |

TINUED.

| *With Bx on Platinum Wire.* | *With SPh on Platinum Wire.* |
| --- | --- |
| As with Carbonate of Soda. | As with Carbonate of Soda. |
| OFl: A small amount of oxide causes the glass to look yellow while hot, colorless when cold. When more of the oxide is present the glass, while hot, appears red, and yellow when cold. A still larger quantity makes the glass dark-red while hot, and dark-yellow when cold.<br>RFl: The glass becomes bottle-green. Treated on Ch with tin it becomes, at first, bottle-green, but afterwards pure vitriol-green. | OFl: When at a certain point of saturation the glass, while hot, appears yellowish-red, and becomes on cooling at first yellow, then greenish, and finally colorless. On a very large addition of oxide it appears, while hot, deep-red, becoming, on cooling, brownish-red, then of a dirty-green color, and finally brownish-red.<br>RFl: A glass containing but little of the oxide suffers no visible change. When more of the oxide is present it is red while hot, and on cooling becomes at first yellow, then greenish, and finally reddish. Treated with tin on Ch the glass on cooling becomes at first green, and finally colorless. |
| As with Carbonate of Soda. | As with Carbonate of Soda. |
| OFl: Easily soluble to a limpid yellow glass which on cooling becomes colorless. If much oxide is present it may be made cloudy by flaming. A still larger addition of oxide causes the bead to become enamel-yellow on cooling.<br>RFl: The glass diffuses itself over the Ch and becomes cloudy. With continued blowing the oxide is reduced to metal, with effervescence, and the glass becomes clear again. | OFl: As with Bx. But to obtain a glass which appears yellow while hot, a large addition of the oxide is required.<br>RFl: On Ch the glass becomes grayish and cloudy. This phenomenon is better observed when tin is added; but the glass can never be made quite opaque. If much of the oxide is present, the Ch becomes coated. |

TABLE II. — CON-

| *Metallic Oxides in Alphabetical Order.* | *On Charcoal alone.* | *With Carbonate of Soda.* |
|---|---|---|
| 13. SESQUIOXIDE OF MANGANESE. $Mn^2O^3$. | OFl: Insoluble. When the temperature is sufficiently high, both the sesquioxide and the peroxide are converted into a reddish-brown powder ($MnO+Mn^2O^3$). RFl: The same effect. | OFl: On platinum wire or foil a very small quantity dissolves to a transparent green mass, which on cooling becomes opaque and bluish-green. RFl: On Ch it cannot be reduced to metal; the Sd passes into the Ch and leaves the protoxide behind. |
| 14. PROTOXIDE OF MERCURY. HgO. | Instantly reduced and volatilized. | Heated in a matrass to redness, it is reduced and vaporized. The vapors condense in the neck of the matrass and form a metallic coating. |
| 15. MOLYBDIC ACID. $MoO^3$. | OFl: Fuses, becomes brown, vaporizes, and deposits on the Ch a yellow Ct, which nearest to the assay is crystalline. On cooling the Ct becomes white, and the crystals colorless. RFl: The greater part of the assay is absorbed by the Ch, and may be reduced to metal at a sufficiently high temperature; the metal is in the shape of a gray powder. | OFl: On platinum wire dissolves with effervescence to a limpid glass, which on cooling becomes milk-white. RFl: Fuses with effervescence. The fused mass is absorbed by the Ch, and part of the acid is reduced to metal which may be obtained as a steel-gray powder. |
| 16. OXIDE OF NICKEL. NiO. | OFl: Not changed. RFl: On Ch reduced to metal; the spongy mass cannot be fused to a globule, but assumes metallic lustre by friction; it is attracted by the magnet. | OFl: Insoluble. RFl: Easily reduced to metal, in the shape of bright, white scales, which are attracted by the magnet. |

TINUED.

| *With Bx on Platinum Wire.* | *With SPh on Platinum Wire.* |
|---|---|
| OFl: Colors very intensively.—The glass, while hot, is violet, on cooling it assumes a reddish tinge. When much manganese is added, the glass becomes quite black and opaque; but the color can be seen when the glass, while soft, is flattened with the forceps.<br>RFl: The glass becomes colorless. If the color was very dark, the phenomenon is best observed on Ch with addition of tin. | OFl: A considerable addition of manganese must be made to produce a colored glass; it then appears, while hot, brownish-violet, and reddish-violet when cold, but never opaque. If the glass contains so small a quantity of manganese that it appears colorless, an addition of nitre will produce the characteristic coloration.<br>RFl: Becomes very soon colorless. |
| | |
| OFl: Dissolved in large quantities to a limpid glass which, while hot, appears yellow, but colorless on cooling. A very large amount of acid causes the glass to appear dark-yellow while hot, and opaline when cold.<br>RFl: A highly saturated bead becomes brown, and opaque when still more acid is present. | OFl: Easily soluble to a limpid glass; if but little of the acid is present it is yellowish-green while hot, but when cold almost colorless. On Ch the glass becomes very dark, and on cooling assumes a beautiful green color.<br>RFl: The glass assumes a very dark, dirty-green color which, on cooling, becomes beautiful bright-green. The same on Ch; tin deepens the color a little. |
| OFl: A small quantity colors the bead violet while hot; when cold, pale reddish-brown. More oxide makes the coloration deeper.<br>RFl: The glass becomes gray and cloudy, or even opaque. With continued blowing the minute particles of reduced metal collect together and the glass becomes colorless. This takes more readily place on Ch, especially when tin is added. The nickel then unites with the tin to a globule. | OFl: Soluble to a reddish glass which, on cooling, becomes yellow. A larger addition causes the glass to appear brownish-red while hot, and reddish-yellow when cold.<br>RFl: On platinum wire not changed. On Ch with tin it becomes, at first, gray and opaque; with continued blowing the nickel becomes reduced, and the glass clear again and colorless. |

TABLE II.—CON-

| *Metallic Oxides in Alphabetical Order.* | *On Charcoal alone.* | *With Carbonate of Soda.* |
| --- | --- | --- |
| 17. BINOXIDE OF OSMIUM. $OsO^2$. | OFl: Converted into osmic acid which, without depositing a Ct, volatilizes with its peculiar pungent odor. RFl: Easily reduced to a dark-brown and infusible metallic powder. | Easily reduced to an infusible metallic powder. |
| 18. PROTOXIDE OF PALLADIUM. PdO. | Reduced at a red-heat; but the metallic particles are infusible. | Insoluble. The Sd passes into the Ch, and leaves the Palladium behind. |
| 19. BINOXIDE OF PLATINUM. $PtO^2$. | Like Palladium. | Like Palladium. |
| 20. PROTOXIDE OF SILVER. AgO. | Easily reduced to metallic silver, which unites to one or more globules. | Instantly reduced. The Sd passes into the Ch, and the metal unites to one or more globules. |
| 21. TELLUROUS ACID. $TeO^2$. | OFl: Fuses, and is reduced with effervescence. The reduced metal becomes instantly vaporized and covers the Ch with tellurous acid; the Ct usually has a red or dark-yellow edge. RFl: As in OFl; the outer flame appears of a bluish-green color. | Soluble, on platinum-wire, to a limpid and colorless glass, which on cooling becomes white. On Ch reduced and volatilized, depositing a Ct of tellurous acid. |
| 22. BINOXIDE OF TIN. $SnO^2$. | OFl: The protoxide burns, like tinder, to binoxide. The binoxide becomes very luminous and appears, while hot, yellowish, but assumes on cooling a dirty-white color. RFl: With a powerful and continued flame it may be reduced to metal. | OFl: On platinum-wire it forms with Sd, with effervescence, an infusible compound. RFl: On Ch reduced to metallic tin. |

TINUED.

| *With Bx on Platinum Wire.* | *With SPh on Platinum Wire.* |
| --- | --- |
| | |
| OFl and RFl: Reduced, but not dissolved; the metallic particles cannot be fused to a globule. | As with Bx. |
| Like Palladium. | Like Palladium. |
| OFl: In part dissolved, and in part reduced. On cooling, the glass becomes opalescent or milk-white, according to the amount of oxide present.<br>RFl: The glass at first becomes gray, but afterwards limpid and colorless. | OFl: Imparts to the bead a yellowish color. When much of the oxide is present, the glass, when cold, is opalescent, and appears yellowish at daylight, reddish at candle-light.<br>RFl: As with Bx. |
| OFl: Soluble to a limpid and colorless glass which, on Ch, becomes gray from reduced metal.<br>RFl: On Ch becomes at first gray, afterwards colorless. The Ch becomes coated with tellurous acid. | As with Borax. |
| OFl: A very small quantity dissolves slowly to a limpid and colorless glass, which remains so on cooling.<br>RFl: From a highly saturated glass a part of the oxide may be reduced on Ch. | OFl: As with Borax.<br>RFl: The glass, containing oxide, suffers no change. |

TABLE II.—CON-

| *Metallic Oxides in Alphabetical Order.* | *On Charcoal alone.* | *With Carbonate of Soda.* |
|---|---|---|
| 23. TITANIC ACID. $TiO^2$. | OFl: Assumes, on heating, a yellow color, and becomes white again on cooling. Suffers no other change. RFl: As in OFl. | OFl: on Ch it dissolves, with effervescence, to a dark-yellow glass, which, on cooling, crystallizes. When cold it is grayish-white. RFl: As in OFl; cannot be reduced to metal. |
| 24. TUNGSTIC ACID. $WO^3$. | OFl: Not changed; at a very high temperature converted into oxide. RFl: Blackens, being converted into oxide, but does not fuse. | OFl: On platinum wire it dissolves to a limpid and deep-yellow glass, which, on cooling, becomes crystalline and opaque, and of white or yellowish color. RFl: With very little Sd on Ch it is reduced to metal; with more Sd it forms a yellow compound of metallic lustre which passes into the Ch |
| 25. SESQUIOXIDE OF URANIUM. $U^2O^3$. | OFl: Infusible; but assumes a dirty yellowish-green color. RFl: Blackens, owing to the formation of protoxide. | OFl: Insoluble. With a certain amount of Sd the mass becomes yellowish-brown, and with more passes into the Ch. RFl: As in OFl; no réduction to metal takes place. |
| 26. VANADIC ACID. $VO^3$. | Fusible. Where it is in contact with the Ch it becomes reduced and passes into the Ch. The rest assumes the lustre and color of graphite. | Unites to a fusible mass which is absorbed by the Ch. |

TINUED.

| *With Bx on Platinum Wire.* | *With SPh on Platinum Wire.* |
|---|---|
| OFl: Easily soluble to a limpid glass which, when containing a large quantity, appears yellow while hot, but becomes colorless on cooling. When containing a very large quantity it is enamel-white when cold.<br>RFl: When containing but little titanic acid the glass becomes yellow; when more, dark-yellow to brown. A saturated glass becomes enamel-blue by flaming. | OFl: Easily dissolved to a limpid glass which, when containing a large quantity, appears yellow while hot, but becomes colorless on cooling.<br>RFl: Appears yellow while hot, but, on cooling, reddens and finally assumes a violet color. If iron is present the glass, on cooling, becomes brownish-red; with tin on Ch the glass becomes violet, unless the amount of iron be very considerable. |
| OFl: Like titanic acid.<br>RFl: A glass, containing but little tungstic acid, is not changed. When more, it becomes yellow and, on cooling, yellowish-brown. On Ch the same reaction is produced with a less saturated bead. Tin deepens the colors. | OFl: Easily dissolved to a limpid and colorless bead, which when highly saturated, appears yellow while hot.<br>RFl: With little blowing the glass appears, while hot, of a dirty green color, blue on cooling; with strong blowing it becomes, on cooling, bluish-green. On Ch with tin, deep green. If iron is present the glass, on cooling, becomes brownish-red; with tin on Ch the glass becomes blue or, if the amount of iron is considerable, green. |
| OFl: Behaves like sesquioxide of iron. When highly saturated the glass may be made enamel-yellow by flaming.<br>RFl: Behaves like sesquioxide of iron. The green bead, when at a certain point of saturation, may be made black by flaming. On Ch with tin it becomes dark-yellow. | OFl: Dissolves to a limpid yellow glass which, on cooling, becomes yellowish-green.<br>RFl: The glass assumes a dirty green color which, on cooling, changes to a fine green. With tin on Ch the color deepens. |
| OFl: Dissolved to a limpid glass which, when the quantity of vanadic acid is small, appears colorless, when larger yellow, and which, on cooling, becomes greenish.<br>RFl: The glass, while hot, appears brownish, and assumes a fine green color on cooling. | OFl: Soluble to a limpid glass which, if sufficient vanadic acid is present, appears dark-yellow while hot, and becomes light-yellow on cooling.<br>RFl: As with Borax. |

TABLE II.—CON-

| *Metallic Oxides in Alphabetical Order.* | *On Charcoal alone.* | *With Carbonate of Soda.* |
|---|---|---|
| 27. OXIDE OF ZINC. ZnO. | OFl: When heated becomes yellow and, on cooling, white again. It fuses not, but becomes very luminous.<br>RFl: Is slowly reduced; the reduced metal becomes rapidly re-oxidized and the oxide deposited on another place of the Ch. | OFl: Insoluble.<br>RFl: On Ch it becomes reduced. The metal vaporizes and coats the Ch with oxide. With a powerful flame the characteristic zinc-flame is sometimes produced. |

TINUED.

| *With Bx on Platinum Wire.* | *With SPh on Platinum Wire.* |
|---|---|
| OFl: Dissolves readily, and in large quantity, to a limpid glass, which appears yellowish while hot; on cooling it is colorless. When much of the oxide is present, the glass may be made enamel-white by flaming; and on a still larger addition it becomes enamel-white on cooling.<br>RFl: The saturated glass becomes at first gray and cloudy, and finally transparent again. On Ch the oxide becomes reduced, the metal vaporizes and coats the Ch with oxide. | As with Borax. |

## TABLE III.—THE METALLIC OXIDES ARRANGED THEY IMPART

### WITH BORAX IN OXYDATION FLAME PRODUCE:

*a.—Colorless Beads.*

| | | |
|---|---|---|
| HOT AND COLD. | Silica, Alumina, Binoxide of Tin, Baryta, Strontia, Lime, Magnesia, Glucina, Yttria, Zirconia, Thoria, Oxide of Lanthanium, Oxide of Silver, Tantalic Acid, Niobic Acid, Tellurous Acid. | when highly saturated opaque (white) by flaming. |
| | Titanic Acid, Tungstic Acid, Molybdic Acid, Oxides of Zinc, Cadmium, Lead, Bismuth, and Antimony. | when feebly saturated. |

*b.—Yellow Beads.*

| | | |
|---|---|---|
| HOT. | Titanic Acid, Tungstic Acid, Oxides of Zinc, and Cadmium. | when highly saturated; on cooling colorless, and cloudy by flaming. |
| | Oxides of Lead, Bismuth, and Antimony. | when highly saturated; on cooling colorless. |
| | Sesquioxides of Cerium, Iron, and Uranium. | when feebly saturated; on cooling colorless. |
| | Sesquioxide of Chromium, when fully saturated; when cold, yellowish-green. | |
| | Vanadic Acid; when cold, pale-green. | |

*c.—Red to Brown Beads.*

| | |
|---|---|
| HOT. | Sesquioxide of Cerium; on cooling yellow, enamel-like by flaming. |
| | Sesquioxide of Iron; on cooling, yellow. |
| | Sesquioxide of Uranium; on cooling yellow, enamel-yellow by flaming. |
| | Sesquioxide of Chromium; on cooling yellowish-green. |
| | Sesquioxide of Iron, containing Manganese; on cooling yellowish-red. |
| COLD. | Oxide of Nickel (reddish-brown to brown); violet while hot. |
| | Sesquioxide of Manganese (violet-red); violet while hot. |
| | Oxide of Nickel, containing Cobalt; violet while hot. |

*d.—Violet Beads (amethyst-colored).*

| | |
|---|---|
| HOT. | Oxide of Nickel; on cooling, reddish-brown to brown. |
| | Sesquioxide of Manganese; on cooling, violet-red. |
| | Oxide of Nickel, containing Cobalt; on cooling, brownish. |

## WITH REFERENCE TO THE COLORS WHICH TO THE FLUXES.

WITH BORAX IN REDUCTION FLAME PRODUCE:

*a.—Colorless Beads.*

| | | |
|---|---|---|
| HOT AND COLD. | Silica, Alumina, Binoxide of Tin. | |
| | Baryta, Strontia, Lime, Magnesia, Glucina, Yttria, Zirconia, Thoria, Oxides of Lanthanium and Cerium, Tantalic Acid. | when highly saturated cloudy by flaming. |
| | Sesquioxide of Manganese; sometimes, on cooling, pale rose-colored. | |
| | Niobic Acid; when feebly saturated. | |
| | Oxides of Silver, Zinc, Cadmium, Lead, Bismuth, Antimony, Nickel, Tellurous Acid. | with strong blowing; with feeble blowing gray. |
| HOT. | Oxide of Copper; when highly saturated on cooling opaque and red. | |

*b.—Yellow to Brown Beads.*

| | |
|---|---|
| HOT. | Titanic Acid (yellow to brown); when highly saturated enamel-blue by flaming. |
| | Tungstic Acid (yellow to dark-yellow); when cold brownish. |
| | Molybdic Acid (brown to opaque). |
| | Vanadic Acid (brownish); green when cold. |

*c.—Blue Beads.*

| | |
|---|---|
| HOT. | Oxide of Cobalt; retains its color on cooling. |

*d.—Green Beads.*

| | |
|---|---|
| HOT AND COLD. | Sesquioxide of Iron (yellowish-green); especially when cold. |
| | Sesquioxide of Uranium (yellowish-green); when highly saturated black by flaming. |
| | Sesquioxide of Chromium (light to dark emerald-green). |
| HOT. | Vanadic Acid; brownish while hot. |

TABLE III.—CON-

WITH BORAX IN OXYDATION FLAME PRODUCE:

*e.—Blue Beads.*

| | |
|---|---|
| HOT. | Oxide of Cobalt; retains its color on cooling. |
| COLD. | Oxide of Copper (when highly saturated greenish-blue); green while hot. |

*f.—Green Beads.*

| | | |
|---|---|---|
| HOT. | Oxide of Copper; when cold blue or greenish-blue. | |
| | Sesquioxide of Iron, containing Cobalt or copper.<br>Oxide of Copper, containing Iron or Nickel. | on cooling the color changes, according to the proportion in which the various oxides are present, to light-green, blue, or yellow. |
| COLD. | Sesquioxide of Chromium, yellowish-green; yellow to red while hot.<br>Vanadic Acid, greenish; yellow while hot. | |

TINUED.

WITH BORAX IN REDUCTION FLAME PRODUCE:

*e.—Gray and Cloudy Beads.*

| | | |
|---|---|---|
| COLD. | Oxides of Silver, Zinc, Cadmium, Lead, Bismuth, Antimony, Nickel, Tellurous Acid. | with feeble blowing; with strong blowing colorless. |
| | Niobic Acid; when highly saturated. | |

*f.—Red and Opaque Beads.*

COLD. { Oxide of Copper, when highly saturated; colorless while hot.

TABLE III.—CON-

WITH SALT OF PHOSPHORUS IN OXYDATION FLAME PRODUCE:

*a.—Colorless Beads.*

| | | |
|---|---|---|
| HOT AND COLD. | Silicic Acid; soluble only in minute quantity. | |
| | Alumina, Binoxide of Tin; soluble with difficulty. | |
| | Baryta, Strontia, Lime, Magnesia, Glucina, Yttria, Zirconia, Thoria, Oxide of Lanthanium, Tellurous Acid. | when highly saturated become opaque by flaming. |
| | Acids of Tantalium, Niobium, Titanium, Tungsten, Antimony; Oxides of Zinc, Cadmium, Lead, Bismuth. | if not too highly saturated. |

*b.—Yellow Beads.*

| | | |
|---|---|---|
| HOT. | Acids of Tantalium, Niobium, Titanium, Tungsten, Antimony; Oxides of Zinc, Cadmium, Lead, Bismuth. | when highly saturated; colorless on cooling. |
| | Oxide of Silver, yellowish; when cold opalescent. | |
| | Sesquioxide of Iron.<br>" " Cerium. | when feebly saturated; on cooling colorless. |
| | " " Uranium; when cold yellowish-green. | |
| | Vanadic Acid, deep-yellow; when cold of a lighter shade. | |
| COLD. | Oxide of Nickel; while hot reddish. | |

*c.—Red Beads.*

| | | |
|---|---|---|
| HOT. | Sesquioxide of Iron.<br>" " Cerium. | when highly saturated; when cold yellow. |
| | Oxide of Nickel, reddish; when cold yellow. | |
| | Sesquioxide of Chromium, reddish; when cold emerald-green. | |

*d.—Violet Beads.*

| | |
|---|---|
| HOT. | Sesquioxide of Manganese, brownish-violet; on cooling pale reddish-violet. |
| | Oxide of Didymium; when cold of a lighter shade. |

*e.—Blue Beads.*

| | |
|---|---|
| HOT. | Oxide of Cobalt; when cold of the same color. |
| COLD. | Oxide of Copper; green while hot. |

*f.—Green Beads.*

| | | |
|---|---|---|
| HOT. | Sesquioxide of Iron, containing Cobalt or Copper.<br>Oxide of Copper, containing Iron or Nickel. | on cooling the color changes, according to the proportion in which the various oxides are present, to light-green, blue, or yellow. |
| | Oxide of Copper; when cold blue or greenish blue. | |
| | Molybdic Acid, yellowish-green; when cold of a lighter shade. | |
| COLD. | Sesquioxide of Uranium, yellowish-green; while hot yellow. | |
| | Sesquioxide of Chromium, emerald-green; while hot reddish. | |

TINUED.

WITH SALT OF PHOSPHORUS IN REDUCTION FLAME PRODUCE:

*a.—Colorless Beads.*

| | | |
|---|---|---|
| HOT AND COLD. | Silica, but slightly soluble. | |
| | Alumina, Binoxide of Tin, soluble with difficulty. | |
| | Baryta, Strontia, Lime, Magnesia, Glucina, Yttria, Zirconia, Thoria, Oxide of Lanthanium. | when highly saturated become opaque by flaming. |
| | Oxides of Didymium, Cerium, Manganese. | |
| | Oxides of Silver, Zinc, Cadmium, Lead, Bismuth. Tantalic Acid, Antimonious Acid, Tellurous Acid. Oxide of Nickel, on Ch. | with continued blowing. |

*b.—Yellow to Red Beads.*

| | | |
|---|---|---|
| HOT. | Sesquioxide of Iron; on cooling greenish, then reddish. | |
| | Titanic Acid, yellow; on cooling violet. | |
| | Vanadic Acid, brownish; when cold emerald-green. | |
| | Titanic Acid containing Iron. Tungstic " " " | yellow; when cold blood-red. |
| | Niobic " " " | brownish-red; when cold deep-yellow. |

*c.—Violet Beads.*

| | |
|---|---|
| COLD. | Niobic Acid, when highly saturated; while hot of a pale dirty-blue color. |
| | Titanic Acid; yellow while hot. |

*d.—Blue Beads.*

| | |
|---|---|
| COLD. | Oxide of Cobalt; of the same color when hot. |
| | Tungstic Acid; while hot brownish. |
| | Niobic Acid, when very highly saturated; while hot of a dirty-blue color. |

*e.—Green Beads.*

| | |
|---|---|
| COLD. | Sesquioxide of Uranium; while hot less bright. |
| | Molybdic Acid; while hot of a dirty-green color. |
| | Vanadic Acid; while hot brownish. |
| | Sesquioxide of Chromium; while hot reddish. |

*f.—Gray and Cloudy Beads.*

| | | |
|---|---|---|
| COLD. | Oxides of Silver, Zinc, Cadmium, Lead, Bismuth, Antimony, Nickel; Tellurous Acid. | takes quickest place on Ch; with continued blowing colorless. |

*g.—Red and Opaque Beads.*

| | |
|---|---|
| COLD. | Oxide of Copper, when highly saturated, or with Tin on Ch. |

# INDEX.

THE END

www.ingramcontent.com/pod-product-compliance
Lightning Source LLC
LaVergne TN
LVHW011233110826
845150LV00006B/1625

*9781425514808*